EXTRA-TERRESTRIAL INTELLIGENCE:
The First Encounter

edited by **James L. Christian**

Ŗ *Prometheus Books*
Buffalo, NY 14215

Published 1976 by Prometheus Books
923 Kensington Avenue, Buffalo, New York 14215

Second Printing

Library of Congress Card Number 76-25328
ISBN 0-87975-064-2

Printed in the United States of America

If we ever establish contact with extra-terrestrial life, it will reveal to us our true place in the universe, and with that comes the beginnings of wisdom.

— Isaac Asimov

$$N = R_* f_p n_e f_l f_i f_c L$$

This formula was developed by Frank Drake and Carl Sagan, astronomers at Cornell University, for calculating the number of technological civilizations in existence in our galaxy with whom we might establish communication.

Contents

ix

Illustrations

Preface

This book is a philosophic enterprise. It is speculative. Fifteen of us—fourteen in original essays written especially for this volume*—have pondered the assumption that we live in a true biocosmos. We have asked ourselves what that assumption might mean if it were actually true.

With the term *biocosmos* I wish to imply a universe that includes life as a natural and inherent event, a cosmos in which life-forms are an inevitable product of the overall processes of cosmic evolution.

These essays *assume* the universality of life. Today this is no arbitrary first principle. An enormous accumulation of evidence from the field of biochemical evolution now tips the scales in favor of the existence of life all over our starry universe.

---

*All of the articles in this anthology are original except for George Abell's "The Search for Life Beyond Earth: A Scientific Update," which has been included to provide background and a discussion of recent developments on scientific aspects of biocosmic activity. Dr. Abell's essay first appeared in *Biology Digest* (March 1976) and is reprinted with permission.

1

We can be virtually certain that higher life-forms are cosmic in scope. Life, sensitivity, perception, intelligence, awareness—all these exist "out there." Life is neither an earth accident nor a finger-of-god disturbance. It cannot be unique to our blue-green planet home. Assuming this to be the case, we have said to ourselves: Let's take another look at the meaning of *human* existence. Let's philosophize again, theologize again. Let's take another look at our laws, our ethics, our minds, our knowledge. Let's look again at death.

This book owes much to Paul Kurtz and Broady Richardson of Prometheus Books: congenial craftsmen, gentle men, and scholars.

I'm grateful in a unique way to the authors of the articles in this volume. A special kind of courage is required to break through the front lines of philosophic speculation to create new frontiers.

Lastly, an appreciative thank you to two men who are present-masters but old hands at conducting tours along the starlanes and sipping tea with starfolk: to Isaac Asimov for going out of his way to help actualize this anthology, and to Ray Bradbury for saying *again,* "Yes! We *will* reach Alpha Centauri!"

JLC

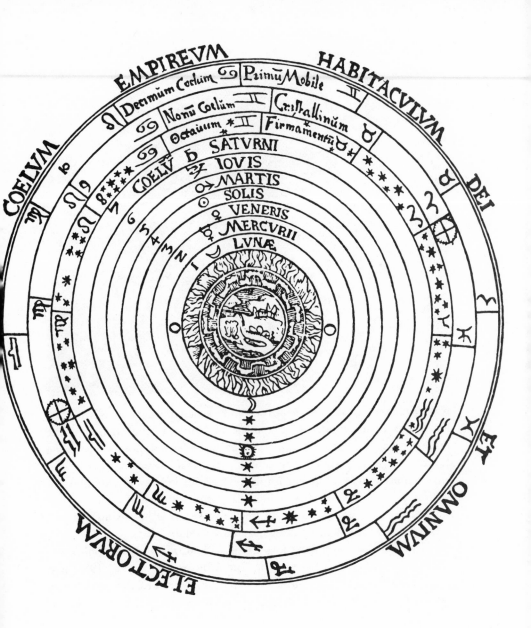

Ray Bradbury and *science fiction* seem like synonyms to many of us. Sci-fi for Bradbury, however, has a specific meaning. Science fiction is the present creation of possible futures. Man-in-time needs *both* a past and a future if he is to see himself in context. Historians *create* our pasts; science-fiction writers *create* our futures. These futures reveal a quintessential reality that, by feedback, enlightens our own time and guides us forward. They scare us, of course, but they also inspire us by giving us a prevision of what man, in greatness, could become.

Ray Bradbury's lifelong love affair with words and metaphors has produced many masterpieces, from *The Martian Chronicles* (1950) and *Dandelion Wine* (1955) to the screenplays for *Moby Dick* and for his own *Fahrenheit 451*, which was translated into a motion picture by François Truffaut.

All told, Bradbury is the author of 350 short stories and 25 books. His recent writings include *Something Wicked This Way Comes* (a novel), *The Halloween Tree* ("a history of death in the world"), *When Elephants Last in the Dooryard Bloomed* (verse), *Long After Midnight* (short stories), and *Pillar of Fire* (plays).

What literary worlds come next? "Perhaps musical lyrics. Perhaps an opera, somewhere down the line. Perhaps an opera called *Leviathan '99.*"

*"Ask the gentle fools and madmen who wrote this book
to sit with you to play and replay the rare, sweet, bitter,
grand games of this strange, weeping, laughing animal
who wakens from monkey dreams to find himself almost
Man."*

Ray Bradbury

Foreword:
The Nefertiti-Tut Express—
A Journey Through
Time, Thought, and Being

In a play of mine, which no one has seen and few have considered, there is
a portion of one act where Heeber Finn, proprietor of an Irish pub, buys a
set of porcelain signs from a traveling flim-flam philosopher.

The signs read: *Stop, Consider, Think, Do.*

Finn puts the signs all about his pub, and soon discovers that the
damn things drive his clients mute, mad, and out into the fogs and mists to
vanish forever. The Irish have *been* philosophizing all along. The last
thing they needed was someone to tell them to Stop or Consider or Think
or, God Forbid, Do. Frozen into immobility by self-consciousness, the talk
dries up, the philosophy drops dead.

I am in a similar situation up front here, playing pub-master,
handing out the brimful drinks to one and all, and beyond that playing
Kabuki prop-master to James Christian, if you don't mind my mixing the
metaphor on the spot.

If I'm not careful, and if James Christian isn't careful, and his pack of

5

philosophers with him, we will make you self-conscious, your mouth will drop open, your brain will drop out, and you will be no good until breakfast or beyond.

The reason for a book on philosophy is, isn't it, to make you more bravely philosophical? Which means, in the end, to dare to be foolish. To, like Mrs. Grundy, blab so you will know what you think, yes?

And if this book doesn't do that for you, it will have failed miserably. And if I get in the way of the book too much, I will deserve to be fired out of hand, with no recompense.

So I come not to stifle but to celebrate. I come to show you how I react to the universe and the mind-meddlers and idea-merchants you find herein.

Let's start with a thought that has often moved across my consciousness late at night. Pretend along with me, please. . . .

Imagine that you have been dead for a year, ten years, a hundred years, a thousand years. The grave and night have taken and kept you in that silence and dark which says nothing and so reveals absolute zero.

In the middle of all this darkness and being alone and bereft of sense, let us imagine that God comes to your still soul and lonely body and says: I will give you one minute more of life. I will restore you to your body and senses for sixty seconds. Out of all the minutes in your life, choose one. I will put you in that minute, and you will be alive again after a hundred years, a thousand years of darkness. Which is it? Choose. Speak. Which minute do you choose?

And the answer is:

Any minute. *Any minute at all!* Oh God, oh Sweet Christ, oh Mystery, give me any minute in all my life.

And the answer further is:

When I lived I didn't know that every minute was special, precious, a gift, a miracle, an incredible thing, an impossible work, an amazing dream.

But now, like Ebenezer Scrooge on Christmas Morn, with snow in the air and the promise of rebirth given, I know what I should have known in my dumb shambles: that it is all a lark, and it is a beauty beyond tears, and also a terror. But I dance about, I become a child. I am the boy who runs for the great bird in the window and I am the man who sends the boy running for that bird, and I am the life that blows in the snowing wind along that street and the bells that sound and say: Live, live, for too soon will your name which is writ on water pass, or your flesh which is shaped in snow melt, or your soul which is inscribed like a breath of vapor on a cold glass pane fade. Run, run, lad, run, down the middle of Christmas at the center of life.

So would we all, in the finale, offered a choice by God, Life-Force, Energy of the Universe—you name it—fall back on simple seconds, incredible minutes.

And if we had a special choice, I suppose, we might simply say: Give me the hour of my birth. Let me come shouting forth to celebrate the Universe which knows not itself but in my birth knows.

So perhaps the moment of birth is very special.

And we would pick that, but beyond that not so much the lustful minutes of our life, for they are tempting, the moments of communion with our friends or our lovers, but in all simplicity such days as I knew in the past year. Let me tell you *my* minute:

Last spring I was taken up to a small hill town in the mountains of California to speak, of all things, to a few score of theological students. Arriving at an altitude of five thousand feet, I was witness to the first snowfall I had seen since I was fourteen years old. I stood in the midst of whiteness, with silence falling all about upon incredible trees and hills, and I wept. From time to time during the day, while speaking to the students, the tears burst from my eyes again and again, and I had to stop and get back my composure with my breath.

That was my moment. That was my minute. But I have had ten thousand times a million such minutes in an entire life, and for each, even as I breathe in sustenance, I breathe out gratitude.

We all pretend at times that we hate life, but life is not what we hate; we hate the way we have used it or not used it, shaped it or wasted it. We hate ourselves, often, for not enjoying the utter simplicity of the miracle.

The long way around to my point being, we may give lip service to the idea that we would be glad to be quit of life, but the facts are otherwise. If God did come to us in the Long Sleep, we would leap to our feet and cry: Give me, oh yes, give me, not a year, not a month, a week, an hour, but if all you have to offer is one minute, I'll take that, yes. Just to be able to breathe again, smell again, taste again, hear, and touch again!

I do not for a moment argue that life is perfection and all beauty. If it were, there would be no need for this book, these writers, this Introduction.

What we must explain to ourselves is how amidst so much light there is dark, with so much health illness must strike, with so much life an equal offering of paradoxical death. These are the things hard to come by, which pose themselves in mirrors but to vanish into basement bins where midnight heaps itself.

To grow into youngness is a blow. To age into sickness is an insult. To die is, if we are not careful, to turn from God's breast, feeling slighted and unloved. The sparrow *asks* to be seen as it falls.

Philosophy must try, as best it can, to turn the sparrows to flights of angels, which, Shakespeare wrote, sing us to our rest. The transmogrification is not easy. It had best be not facile. It should not hypocritize us. Our funeral sermon, like our birth announcement, should stick to the facts, if we can find them, and celebrate without being vulgar.

It is not easy. But we all go the same way home. All stroll together in some similar kind of weather, as the old song moves.

My dear friend and animator, Chuck Jones, one day recently telephoned me in high spirits. He is always telephoning me in high spirits which makes our friendship extraordinary, for often one's friends call you only when they are in low despair and need you to hype them. Chuck always telephones with news fresh out of a slang dictionary or a thesaurus he is reading as if it were *War and Peace* and twice as wonderful.

This particular day, the animator of Wylie Coyote and the Roadrunner asked me what I knew about the stoking facilities on the trans-Egyptian railways back in Victoria's late time. I pleaded Arabian ignorance.

"Ah, ha!" cried Jones. "Did you know that when they ran out of fuel, the fireman aboard the locomotive leaped off into the nearest graveyard and fetched ancient mummies aboard and shoved them in the fire as fuel?"

"I didn't know *that*!" I cried.

"*Now* you know it," said Chuck, and hung up.

I sat down and wrote a poem titled "The Nefertiti-Tut Express." Which is the long way around to making a point about life, fuel, philosophy, myself, Jim Christian, and this book, with its new findings and insights about biochemical evolution that tell us we belong to a cosmos of being.

In life, any fuel is good fuel if it makes the engine run.

What works for locomotives applies to human beings.

If Nietzsche stokes your furnace, hire him.

If Schopenhauer is your kindling, become a Schopenhauer-fire-eater.

And if the Stoic philosophers or the authors of this book are sitting around in their mummy-rags begging to be popped into the locomotive bin to become proper torches to light your way, pop them, light them, use them.

So along through life we take on a passenger list of most peculiar mummies and shove them in the tinderbox and hope that they move the ponderous machinery of our dumb minds half a revolution round or half a mile ahead in the prevailing dark.

That's pretty much the story of most of our lives. We dare not call ourselves philosophers, and yet we all are, aren't we, or we could not get out of bed in the morning.

Unless you plan, the night before, to be alive at dawn, you will not stir. The most precious hour is that one just before sleep when we damn the Past, consider the immediately gone Present, and idly dream a better day Tomorrow. When men and women do all three things in one, they are automatically philosophers with a small *p*. Each of us has a love of knowledge even if it could be printed on a thumbsized miniature book and shoved in our ear. Ask any man at breakfast how the world should be run, and he'll tell you. Ask any woman what men are made of and how she plans to reshape the new model, and she'll tell you that.

We have opinions, oh God, we never lack for them.

And native or brute wisdom or natural thought or plain country-meadow wits or intuition or whatever you wish to call it, it is simple green-grass front-yard philosophy, that practical yardstick by which we measure an inch and claim it is a mile, and, since death may not arrive for an hour, with it we have time to compare science and religion and find them *both* wanting, one last time!

There's hardly a question in this book that a man or woman, on a front porch on a summer night, hasn't already answered. Just ask them. Sit back, light a cigar or cigarette, and watch the flow. Once off, your average human being can churn forth Great Ideas fourteen to the dozen late into the night.

Money? They have a theory.

Sex? In one harem and out the other.

Heaven and Hell? They've just come back. Sell you a Guide Book, cheap.

God? They've been on intimate terms with Him for years. Behave yourself and they'll introduce you.

So there you have it. We are all teachers, like it or not, and all philosophers, pull back as we might, or return as we must; for in the long haul,
push, pull, tumult, quell, triumph, despair, the dance or the slog from
dreaming womb to dreamless tomb we must tell each other the way. Youth
hopes it knows, age *knows* it knows; and so they share out an ignorance
and strike a match and somewhere find a torch and fall off a cliff only to
hang by their nails and survive.

Survive may well be the name of the game, but celebrate is another to
which, in our own small ways, we give equal time. There are such days in
the world and we know them well. When we were child-philosophers we
Rorschach-tested the clouds and found wonder, no matter which hill we
lay on, no matter what cirrus or alto-stratus passed in the high places. It
helped, of course, if we lay with loving friends in the years before our genetics invented sex and complicated friendships. But there we were, breathing in time, breathing out love, giving names to everything, be it foxglove,
thyme, or maidenhair fern.

Once we were *all* Gerard Manley Hopkins, for, think on it, isn't
Gerard Manley Hopkins all of us snug/compact, done up like clam in
shell, nautilus in chambered castle, so incredible that God speaks from
every whorl?

We as children, of course, love knowledge, love wisdom, and gather to
it as to gumball machines or mirrors, to be fed or to reflect.

Witness that same front porch again on summer nights, where older
people gabble and maunder and yammer about the Past and how it adds
up, or the Future and what's its horizon. You will see children creep onto
the porch as stealthily as burglars to plant themselves amongst the potted
ferns to hide and lie down and, yes, listen to what life Is All About. And if
the grownups slow down the child whispers, More, Grandpa, or, Hey,
Grams, what happened in 1901, or, Dad, what was it like, what did you
think, harvesting the western fields in 1898, or sleeping in a free jail bed in
1899, or riding the rods in 1900? what, what, what?

And the moments of Truth occur, one by one, voice by voice, life by
life, memory by memory, and this is how it was, they say, and this is what I
thought of what it was, they say, and this is how we hope it will be with
you, they say. Think these thoughts. Love this way. Know these times.
Remember your future as I forget my Past.

They may not be the best poets with the grandest language, but if you
listen closely you might hear something almost like this:

If you wish you can speak of us and God as if we were symbiotic, which is simply a complicated way of saying interacting pals. Just as flowers cannot live without bees, so bees cannot live without flowers; they are a traveling beauty of symbiosis. If God is a rhino, we are those birds that pick the rhino's teeth and use his spine for a landing platform. God needs us to rekindle Himself. The Life Force needs us to speak in tongues. The Universe needs us to see how miraculous it is. The void between the stars needs us to fill it with our shouts of exaltation. We are the hearth on which God basks. We are the flesh in which God expresses Himself, speaking one minute dumb, the next bright, sometimes shambling, sometimes arrow straight.

Listen! All that said on a front porch? Darn tootin'! And more! We are the energy and force and weight of the entire universal country that spreads in stars beyond the power to grasp, hold, imagine. We are the sum total of the magnificent, sad, strange, wild, lovely, terrible, beautiful experiment in which God slides to disaster but saves Himself in flesh. We *are* that stuff, deny it as we will, doubt it as we often do, pray to it as we must.

A final metaphor, if you please.

Imagine a Future Room with forty men and women seated with alternately empty chairs between. Eighty chairs in all, but only forty occupied. It is a robot's banquet in the year 1999, and I have been invited.

I enter and am greeted with a chorus of voices. The men and women at the tables raise their glasses and call out:

"Here, sit here, no here, here!"

And I sit now with Aristotle, now with Emily Dickinson, now with Schopenhauer, in a great feasting of thoughts and a banqueting of words.

"Dear Mr. Bradbury!"

Plato seizes my hand.

"Sir," I say. "How goes it with your 'Republic'?"

"Superb! Fine. Except—"

And he tells me. And I listen.

And I rise to change seats and speak to Sara Teasdale or Albert Camus. I rise and go now with William Butler Yeats. I take tiffin with the somewhat darker philosophies of Shakespeare or the lighter but still deep philosophies of Shaw.

So you may, in this book, move around the endless table, breaking your fast with splendid works, meeting and basking with talented people just as I might someday meet the people I have mentioned reborn in robots to outlast time.

Sit with Kazantzakis and he might shout: God cries out to be saved. We shall save Him. And in symbiotic cries of relief, He will then save us. For we are One.

Make up your own table, guest list, menu.

Sit Gerard Manley Hopkins there and he will perhaps doodle on the linen napkin: What I do is me. For *that* I came.

Ask the gentle fools and madmen who wrote this book to sit with you to play and replay the rare, sweet, bitter, grand games of this strange, weeping, laughing animal who wakens from monkey dreams to find himself almost Man.

If I may be allowed the liberty, this book, and the stuffs in it, say to me: Man will not die . . . ever. Hate as we often hate ourselves, yet we love ourselves more and the gift given us to live one time in a Universe as brutal, stunning, and nightmarish as it is beautiful, beautiful, Oh, then again beautiful beyond our powers to say.

Students, teachers, readers, philosopher-friends, wander the halls and the rooms of this book.

Plato calls you to breakfast.

I would not, if I were you, be late.

We Tellurians have been dreaming for a long time of a people-populated universe. Isaac Asimov notes in his article that this dream didn't make much sense until Galileo came along with his telescope. It wasn't logical to imagine other worlds inhabited by other beings until we knew that there were other worlds to be inhabited. But the logic of the mind only occasionally inhibits the logic of the heart, and man has kept right on conjuring up cosmic creatures to keep him company.

We dream because we are lonely. Yet part of the dream is that our dreams may not be in vain.

For some of us those space photos of a fragile earth floating silently against a black emptiness conveyed a terrible sadness. What if we actually were alone. . . .

At last we have sufficient—though still indirect—scientific evidence to allow us to work with realistic dreams.

Besides having edited this anthology, James Christian is the author of a widely used textbook, *Philosophy: An Introduction to the Art of Wondering.* He teaches philosophy at Santa Ana College in California.

During the past two decades scientists working in the field of biochemical evolution have given us a new world-view. Quietly, they have begun to answer the momentous question of how life begins. But they have revealed to us something more, something that will eventually alter the foundations of human existence. . . .

James Christian

The Story of Life: Earth's Four-Billion-Year Beginning

In 1904 my father watched Orville Wright singe the treetops with a for-ever-trip of about a mile; and he also watched with disbelief—since he is not yet immune to miracles—the 1969 moon landing of the Eagle. Later he marveled at the first free-floating space dwellers aboard Skylab. On warm summer nights he still counts the manmade satellites streaking silently across the darkened Arizona sky.

And perhaps I learned it from him: I find that I become frustrated when the miracles of life are taken for granted.

On a wide-bodied jet gliding through the thin skies at six hundred miles per hour, we humans actually fly; but across the aisle a five-year-old boy props himself against his armrests and, trying to cope with boredom, slurps a coke and toys with his seat belt—oblivious to the miracle of which he's a part.

Or we stare through joyous tears at the fiery lift-off of a moon-bound Saturn rocket and then, just hours later, a new species of man writes dusty

15

signatures on the surface of the lunar terrain; but our teen-age children, having been born in the "space age," wonder what all the fuss is about.

On some clear night, stand alone under the stars with your eyes closed. Imagine very hard that you don't exist—all senses shut down, consciousness turned off. Then suddenly open your eyes and breathe in the millions of Milky Way stars. Your window onto the universe has just opened, dramatically. From your own personal pint-sized space station you look out onto all that is. *You perceive. You are aware. You are alive.* And for a few short earth-years you and I are privileged to peer through that perceptual window.

"I have the key to happiness," wrote Eugene Ionesco. "Remember to be profoundly, totally conscious that you are." But then, speaking for us all, he added: "I myself, sorry to say, hardly ever use this key. I keep losing it."

The true miracles of any age lie not in the action-events that move our emotions but in the insights of the human mind.

Of course it's more joyous to sail the skies than to scan the schematics of a jumbo jet; more fantastic to watch a lunar buggy skirt the craters around Hadley Rille than to ponder the blueprint of a Saturn V and to know, in theory, that it can propel human beings to our neighboring demi-planet. There's no substitute for the adventure of doing.

Yet, in an important sense the *idea* that the mind has labored to develop is the prior miracle. This is no mere reiteration of a Platonic doctrine that the idea is more real than the object produced from it; or that the contemplative condition is more virtuous in some way than the life of action. Rather it's a recognition of the fact that our contribution to the future of our species lies in the knowledge-gains, the insights, and the breakthroughs that take place in the ongoing accumulation of wisdom and awareness.

The philosophers of history speak in unison in reminding us that pyramids and Parthenons will crumble, that nations will rise and fall, and that whole civilizations will color our paper globes and then vanish forever. The legacy that endures is one of human understanding and human character.

In the history of human consciousness intellectual breakthroughs are the major beachheads by which the human race progresses. Man's mental sojourn has been brief, and we can be proud of the pace with which we have broken nature's codes. Still, these quantum leaps are too rare, and some of us stand disbelieving if breakthroughs go unnoticed and if our excitement is not shared by others.

During the past twenty-five years scientists working in the field of biochemical evolution have made this sort of breakthrough. They have laid the empirical foundations for answering one of the persistent, stubborn questions raised by the mind of man: *What is the origin of life?* And, as so often happens after our mind is fortified with facts and begins to make its way into a solution, we look back and say, "Of course! *Now I understand.* How could it have been otherwise?"

This sort of philosophical question is not one of the common garden variety. It's one of the four great etiological questions that have baffled and irritated man's understanding. These four questions are: What is the origin (and nature) of *life,* of *man,* of *matter,* and of *the universe?* These **problems have always appeared to be tangled Gordian knots that the** human mind could not cut.

We had every reason to feel stymied. Not one of these questions could be answered until enormous amounts of scientific data had been gathered and correlated. Two developments had to take place in Western thought. First, the sciences—physics, chemistry, biology, etc., and their numerous subdivisions—had to be born. They had to define their boundaries, specialize in depth, and reach advanced levels of data accumulation and theory within their ever-narrowing fields.

Then, in due time, an interdisciplinary countermovement had to begin. Scientists in the specializing disciplines recognized that they were laboring on different aspects of the same problems and realized that progress could be accelerated if they reestablished communication and cross-fed information. Not a little of the impetus toward coordination has come recently from crisis problems (e.g., pollution, food shortage, abortion and the "right to die" cases in our courts, and cancer research) whose solutions required interdisciplinary insight.

Before the stages of specialization and reunification were reached by Western science, the answers to these etiological questions had to be mythical. Man was caught in the perennial human predicament—a condition we all can appreciate. He couldn't stop asking questions, yet he possessed no factual knowledge that would lead him to the understandable answers his whole being demanded. This predicament is always accommodated by an interim mythical answer that is acceptable only until empirical data become available and a rational understanding can be achieved.

These questions were dispatched with celebrated pragmatic myths. *Whence man?* He was created from white, red, and brown clay; or he was sculpted from rock, or carved from wood, or assembled from pine bark, turquoise chips, and crow feathers. He was created by Tiki, Juok, i Kombengi, Yahweh, or one of a thousand other anthropomorphic man-makers.

And *whence matter? Whence the universe?* On these two etiological puzzles little could be said. Matter apparently exists eternally, shaped into familiar forms by a Demiurge; or a supernatural X-factor created everything in the universe *ex nihilo*, out of nothing. Even when critical minds tried to think carefully about the cosmos, there was still very nearly nothing to be thought. Aristotle is typical. His notion of an Unmoved Mover—the First Domino in an infinite series—was the only near-logical answer to the problem that he could come up with. Virtually no progress was made on any of these problems until the twentieth century.

As of 1976, however, two of these wild dragons—the origins of life and of man—have been domesticated: the rough outlines of an empirical answer are now clear. And the other two—the origins of matter and the universe—have been tamed: viable questions and empirical models are beginning to be developed.

The theory of chemical biogenesis has given us a new world view. We don't live in the same universe we lived in before. The difference is that between a universe of nonliving matter and a universe of life. This fact makes possible a whole new way of perceiving the heavens and relating to them. We have therefore reached a point in time when we can beneficially initiate a re-examination of the nature and meaning of human existence in this new biocosmic context.

From scientists working in biochemical evolution have come three hypotheses that are germane to the speculations in this book.

First, they have begun to answer the question of how life evolves. The opening step was to develop *theoretical models* of how life might have been synthesized on earth during the first phases of our planet's existence. These theories were based on informed guesses regarding the conditions that prevailed at that time: composition of the atmosphere, seismic and volcanic upheavals that shook continents, electrical storms and ultraviolet radiation that relentlessly bombarded the planet's surface, and other cataclysmic events that provided the setting, the substances, and the sources of energy out of which life could evolve. These were brilliant beginnings, but the models were purely theoretical.

Then, during the last quarter-century—specifically, since Stanley Miller's landmark experiments in 1953—scientists have been engaged in research designed to develop *empirical evidence* that, by extrapolation, would give us the first faint insights into the primeval processes that produced life on earth.

We now have a general understanding of how living organisms evolved from inorganic constituents. The living evolved from the non-living. That is, fully living, replicating organisms are composed of, and developed out of, the basic elements and simple organic compounds of our universe. This emergence of life on earth occurred some four billion years ago, when environmental conditions were quite different from those of today. Whether the process took hundreds of millions of years or was a relatively rapid sequence of events is still unknown.

There is a high probability that this is the universal pattern by which life evolves. All other hypotheses, including hylozoism, spontaneous generation, and the untold variety of etiological myths have been left far behind by this empirical biogenetic model.

Second, we can therefore infer that life will have evolved as a normal, natural event anywhere in the universe where the environmental conditions permit the synthesis of simple living organisms.

What reasons do we have for concluding that life may be universal? We think this must be the case because the chemicals necessary for life are found throughout the universe. No matter what heavenly object we train our spectroscopic sights on—the sun, stars, novas, nebulas, dust clouds, galaxies—we find all the familiar elements: hydrogen, helium, nitrogen, carbon, sulfur, and the rest of the elemental family found on earth. Whether we are looking nine light-minutes to our sun or eight billion light-years to Galaxy 3C 123, no strange, unidentifiable element ever occurs. The elemental constituents of living organisms are cosmic. Truly, "we are the stuff of which stars are made."

During the past dozen years or so not only the basic elements but a wide variety of organic compounds heretofore unsuspected have been detected in interstellar space. The list of known compounds—vital to life processes as we know them and on earth often produced by life processes—now contains more than thirty familiar names. Among them are such common compounds as cyanamide, formic acid, acetonitrile, methyl formate, acetaldehyde, formaldehyde, methyl alcohol, and ethyl alcohol (which has given rise to droll comments about "cosmic cocktails").

Therefore the basic elements and organic combinations of these elements necessary for life are universal. The inference seems clear: Since all the required ingredients are abundantly available throughout the universe, life can and will have evolved, given sufficient time, wherever a congenial environment should happen to exist. "The alphabet of life is obviously extremely simple," notes Cyril Ponnamperuma; "a handful of chemicals are responsible for the vast variety we see in the entire biosphere."

Third, if the evolutionary processes known on earth—that is, change through mutation of the gene code, adaptation to environmental demands, and the competitive struggle for survival—if these mechanisms are universal—and we have good theoretical reasons for believing that they are—then in countless cosmic ecosystems there will be some species that have evolved into complex forms and attained advanced stages of perception, consciousness, awareness, and abstract rational capability plus qualities of consciousness quite beyond what man has so far developed or can at present imagine.

In other words, very advanced higher life-forms will have evolved. There is no valid rationale by which we can continue to hold that, as an intelligent organism, man is unique in the universe, or that the cosmos turns and struggles for his singular benefit and salvation. It is a virtual certainty that extraterrestrial creatures who are intelligent, aware, and sagacious do in fact share our universe. Whether they will resemble the fantasy forms of man's science fiction or even the strange configuration of Homo sapiens himself is a datum we will continue to worry about until contact begins to provide the answers.

However that anxious question is eventually answered, what is certain is that we humans live in a true biocosmos. We are part of a living universe composed not merely of lower life-forms on the order of algae, rotifers, lichens, and plants. In terms of man's search for relatedness and meaning, it would not be enough to know that only low life-forms exist. *The meaning of human existence changes because we belong to a biocosmos of intelligence.* High life-forms exist that, judged by specific qualities of perception and consciousness, are our peers and our superiors. It is with this recognition that the phrase "we are not alone" takes on meaning.

When did life begin on planet earth?

The solar system was born about five billion years ago, and by a half-billion years later our planet had condensed into a round hot ball, still violent and inhospitable, washed with storms and tides, but gradually cooling and stabilizing. The generally accepted birthdate for Sol's third planet is 4.6 billion years ago.

Then, about three billion years ago, a momentous change occurred in the earth's atmosphere: large amounts of oxygen began to be produced. Small amounts of oxygen may have already come into existence from the ultraviolet dissociation of water molecules and from the beginnings of green-plant photosynthesis. But as the first plantlike organisms began to produce their own organic nourishment and increased their photosynthetic activity, ever greater quantities of oxygen were released. The earlier reducing atmosphere was suffused with an abundance of oxygen, and this event altered the basic conditions for life on earth. It produced an ozone shield that filtered out destructive ultraviolet radiation and introduced a greenhouse effect. As more free oxygen was produced, conditions conducive to the beginning of animal evolution came into being. It is quite possible that this event triggered the explosion of the evolution of higher species at the beginning of the Cambrian period.

We know from the fossil record that the first hard-shelled animals evolved during the late Precambrian era about 700 million years ago and continued to diversify in a proliferation of species throughout the Cambrian period (beginning about 600 million years ago). These hard-shelled animals conveniently left a fossil record. However, we can be sure that soft-bodied animals, including countless species of single-celled organisms, had by this time passed through a long history of evolutionary development. To date they have eluded fossil hunters.

The oldest known forms of life are algalike cells that lived some 3.5 billion years ago. It appears that these cells were capable of photosynthesis, and chances are that this process had been going on for some time.

When, then, did life begin on earth? All we can say at present is that living organisms evolved some time between 4.6 billion years ago, when the earth was formed, and 3.5 billion years ago, the date of our earliest record of microfossil life. During that one-billion-year period, some wonderful and incredible events were taking place.

The beginnings of the science of biochemical evolution are associated with the work of three men.

In 1922 a Russian biochemist, Alexander Oparin, delivered before a group of scientists in Moscow a paper outlining his theory of biogenesis. Two years later he published his thoughts in a booklet entitled *The Origin of Life*. In 1928 the English biologist J. B. S. Haldane published a technical paper with a similar line of thinking. Both men developed coherent theoretical models from their knowledge of physics and biochemistry, but there was as yet no empirical evidence to support their speculations. Then in 1953 the American biochemist Stanley Miller performed experiments that began to lay empirical foundations for an understanding of how life evolves.

Oparin had theorized that in the earth's early stages a variety of organic compounds had already developed out of inorganic materials. He described theoretically how these compounds could develop into the first prevital organisms and then into living things. As the crust of the earth began to form and the temperature of the atmosphere dropped below a thousand degrees centigrade, a variety of chemical reactions took place. Torrential rains poured down upon the young planet, accompanied by constant discharges of lightning. Hot pools of water formed containing organic compounds that washed down from the atmosphere. Most important, Oparin thought, were the carbon bonds that formed in ever-larger molecular chains. Fatty acids, sugars, and tannins could have formed in this way. Eventually amino acids could have been naturally synthesized, and amino acids are the basic constituents of proteins.

Thus, during the first phase of earth's history—perhaps a billion years long—mixtures of hydrocarbons, nitrogen, hydrogen, and ammonia were continually producing an endless variety of organic compounds that formed complex molecules that became the building-blocks of living cells. At this stage the earth was covered with what Haldane called a "hot dilute soup" in which these prebiotic reactions were taking place. With the synthesis of proteins the first steps had been taken toward the development of life.

The first empirical evidence that Oparin's theory might be correct came from Miller's experiments at the University of Chicago. Into a simple glass apparatus Miller introduced methane, ammonia, and hydrogen. The one essential ingredient of all life—carbon—was there in the methane (CH_4). As these chemicals mixed with vapor from boiling water and passed through glass tubes, they flowed across two tungsten electrodes generating a continuous electric spark. All this was designed to simulate hypothetical primitive-earth conditions—the circulating gases represented the early atmosphere, the flask of boiling liquids represented the young oceans. The experiment ran continuously for a week. At the end of that time, the gases were pumped out and the brownish liquids were analyzed. He found that a variety of organic compounds had formed, along with several amino acids. Miller notes that one of the unexpected results was that "the major products were not themselves a random selection of organic compounds but included a surprising number of substances that occur in living organisms."

Since 1953 Miller and other scientists have added a vast amount of supporting data. Various gaseous mixtures have been tried along with other forms of energy, and in every case biochemically significant molecules were synthesized.

Many specific ingredients essential to living things have now been formed in the laboratory under possible primitive-earth conditions. These include the creation of carbon chains, polypeptides, and ATP, a catalytic enzyme that supplies the basic source of metabolic energy in living systems.

Significant also is the laboratory creation of porphyrins, molecules that function like plants in being able to utilize light to store energy—a primitive kind of photosynthesis. Miller believes that "almost certainly they became important for the metabolic processes leading to ATP synthesis early in the evolution of life." This supports the suggestion that photosynthesizing cells were among the earliest forms of life.

Another biochemist, Sidney Fox, synthesized microspheres, which he named "proteinoids" because they looked and behaved so much like living protein cells. They possessed a double-layered surface analogous to a membrane, and they carried on a kind of internal enzyme activity. Like living cells, they were sufficiently stable to permit sectioning and staining for microscopic examination. Most significantly, they performed a sort of reproduction. When left standing for a week in liquid the spheres formed small attached minispheres or "buds" that could be split off from the parent-spheres. These would proceed to grow to the size of the original cells by the ingestion of selected substances, and then stop growing. In a few days these "offspring" would produce their own "buds" and replication would continue.

The most crucial achievement has been the synthesis of the purines and pyrimidines. The five strategic amino acids have been formed: adenine and guanine (purines) and cytosine, uracil, and thymine (pyrimidines). These nucleic acids form the DNA gene codes and the RNA protein-translation mechanisms for all living things on earth. In turn these amino acids have been joined with sugars and phosphates to create nucleotides, the basic links in the genetic code. Furthermore, it has been shown that these amino acids could have formed with comparative ease on the primitive earth.

In 1975 the first laboratory synthesis of a complete mammalian gene was accomplished. It was a relatively simple hemoglobin gene composed of 650 nucleotides. The report of the event in *Science News* carried the comment: "It's hard to believe that in a swift quarter-century, biologists have made the quantum leap from the identification of hereditary material to its synthesis. Yet that is precisely what has happened."

While such phrases sound like descriptions of living organisms, it must be emphasized that they are not living—yet. All these achievements are only stepping-stones toward the complexity required for the laboratory synthesis of true living organisms.

One of the world's foremost biochemists, Cyril Ponnamperuma, is at once hopeful and realistic in assessing the future of biochemical evolution."There is no reason to doubt that we shall rediscover, one by one, the essential conditions which once determined, and directed, the course of chemical evolution. We may even reproduce the intermediate steps in the laboratory. Looking back on the biochemical understanding gained during the span of one human generation, we have the right to be quite optimistic. In contrast to unconscious nature which had to spend billions of years for the creation of life, conscious nature has a purpose and knows the outcome."

Much labor lies ahead before those steps in the long journey are complete and the story of life can be told from the beginning.

Why does it mean so much that we can now say, "We *understand* how life evolves"?

The theory of chemical biogenesis will become a first-magnitude *field theory*. It will be comparable in its effects to the Pythagorean discovery that mathematics is the key to understanding physics, to Darwin's theory of biological evolution, and to Freud's conceptualization of a subconscious inner world.

Such field theories lead to new and better interpretations of a wide range of phenomena. They are the bright lights of man's intellectual history. They illuminate vast areas of darkness and chase away myths. Because general principles can be developed from them, they lead to solutions of numerous other heretofore unyielding problems. The Pythagorean insight, for instance, opened the door to an interpretation of the entire range of physical phenomena from particles to pulsars. Similarly, Darwin's insight provided a clear theoretical understanding of the relationship of all living things to one another; and this in turn permitted the classification of all life-forms on earth and an understanding of how they evolved through time from simpler organisms. It laid to rest our queries and qualms about the puzzling structural and functional similarities shared by the higher vertebrates. It also accounted for the origin and nature of human races and all their physiological differences.

In just this way the theory of the chemical origin of life will eventually help us solve whole classes of problems that occur on the border line between the nonliving and the living, especially problems involving the pathology of genes and cells. In due time we will have clear definitions for distinguishing the living from the nonliving. We may even gain an understanding of what life per se actually is—or whether "life per se" is a viable notion at all.

The critical consequences will lie in the almost unlimited control that man will possess in laboratory experiments as well as in real-life conditions. The manipulation of gene codes will permit a far more fundamental kind of scientific activity than has previously existed. In all probability we will soon know precisely what arrangements of nucleotides produce specific characteristics; this knowledge in turn will enable scientists to synthesize DNA linkages to produce any desired replicating template. Science fiction will again become fact: man will bring to life creatures that he has designed on the drawing board.

Ethical considerations raised by the creation and control of life will be complex. Having to face new kinds of ultimate concern will stretch our moral tolerances to their limit and, hopefully, force us onto new planes of ethical awareness. The beneficial consequences for life will be enormous, but our wariness is justified. Although the statement is largely a play on words, in the minds of some, man will have become a god—the designer and creator of life. Will he be a wise and competent draftsman, or might he be by nature a caricaturist? If the demonic in his subconscious splashes out onto the drawing board, then we have reason to fear. In any case, we will be compelled to monitor biogenetic activities and set parameters within which the "life designer" must work.

The cosmic implications of the theory of chemical biogenesis are revolutionary. The fact that life will evolve anywhere in the entire universe where congenial conditions exist will eventually become the nucleus of a new world view. Yet at present the discoveries of biochemical evolution have not made a profound impact. "This tremendous event is still on its way, still wandering"—to misquote Nietzsche. "It has not yet reached the ears of man."

Parochial theories of the origin of life have been rendered irrelevant. Earth life is not the accidental result of a near collision with a passing star. Nor does any vital impulse-to-life or finger-of-god intervention seem required by our present biogenetic picture. Our religious etiologies have suffered the fate of all myths.

Field theories such as that of chemical biogenesis spread out like dye to color the interpretation of vast areas of human experience. They become the building blocks of new world views. They are not merely abstract principles written into dusty pages, to be shelved to await the eye of the historian. They lay the foundations upon which future generations will build a greater awareness and give meaning to their lives.

The theory of life's chemical origins will lift the burden of belief from many who still feel they should accept traditional etiological myths. The question of life's origins has at last been removed from the jurisdiction of the mystical and the supernatural.

The insight that life is a natural occurrence will be painful for some, a welcome relief for others. For the individual who feels that life processes must necessarily be associated with the divine, further labor will be required to resolve apparent contradictions. For others, the mythical etiologies that are rooted in man's prescientific past have lost their power because, for them, understanding takes precedence over belief. It's a breath of clean air to feel no longer the pressure to accept answers that never carried with them the promise of understanding.

When myths are released from their beleaguered burden of meeting man's need for answers, they can be seen and appreciated in new ways. This is an important step in the history of man's gradual demythologizing of his world views. During the long prescientific phase of his development, myths were man's interim life-sustaining accounts of the phenomena of experience until scientific theory could provide his mind with better tools for understanding those phenomena. After the advent of empirical models, mythical accounts can then be appreciated for their life-serving functions, for their revelations of the human psyche, and for their charm and literary beauty.

When evaluating insights the sheer joy of understanding should not be minimized. For those who struggle for light, insight is its own reward. It's the intrinsic satisfaction of basking in the solution to a problem one has labored hard to unravel. We may be stumped by a simple algebra problem, for instance, a problem that has no practical application at all; yet arriving finally at that slippery solution feels fantastic.

If the understanding involves only some cranky problem that one single mind has heretofore been unable to fathom, then the gratification is private and spills over only to those who care. But if one is moving within the perspective of the history of human understanding, then the excitement is experienced vicariously for all mankind. It is mankind's insight, and with old Euclid the shout of "Eureka, I have found the answer!" is proclaimed for all men.

I occasionally wander over inshore reefs and explore the tidepools. Anemones, nudibranchs, and a few purple sea urchins lend color to each tiny basin of water. When I turn over rocks brittle-stars scamper to shelter, and hermit crabs, dragging their adopted homes with them, scurry away from my fingers. Limpets and chitons are glued to the smooth sides of the larger wave-washed boulders, waiting for the returning tide to bathe and nourish them. In a small shell-rimmed cavern I find a young octopus, withdrawn from the world, trying to remain below water level. Clusters of mussels cover the rock surfaces of the intertidal zone.

All these creatures stay within the boundaries of their own niches, living in a complex relationship with one another.

I wonder about life—how it evolved, how it has adapted. I wonder about each creature's singular struggle to stay alive in its tiny world—to eat but not be eaten. I wonder about species pushing against other species, preying on one another, or equally often, living together in delicate symbiotic systems. And I wonder if we truly understand all the basic mechanisms of evolvement, or whether, in the larger story, something might be happening that we haven't yet fully captured in our conceptual formulas. The microbiologist René Dubos has long had the feeling "that life entails the operation of some principle of nature which is as yet ill defined."

My reflections wander away from the tidepools at my feet and expand to envision our entire blue-marble planet as a round, watery tidepool teeming with life.

Since the time that living organisms emerged from the "hot thin soup" the proliferation of species on our planet staggers the imagination. Taxonomists—those ever-patient classifiers—have so far discovered, ordered, and described about 1.5 million species of living organisms, and some ten thousand new species are added to the list annually.

To date they have recognized some eighty-six hundred species of birds and nearly twenty-five thousand species of fishes. Their catalogs list nearly 125 thousand other marine organisms. Yet scientists estimate that at least a third of our planet's fishes are still unknown.

Some three-quarters of a million species of insects are recorded, and six to seven thousand new ones are added yearly. The higher vertebrates are mostly accounted for, although several new mammals have recently been discovered. The invertebrates are relatively unknown, especially the mites, nematods, worms, and parasites, which total hundreds of thousands.

About a half-million higher plants have been classified, but taxonomists estimate that a quarter-million species are still unaccounted for, especially in the tropical zones. Lower plants, such as fungi, have hardly been touched.

It is estimated that on our planet perhaps ten million species of organisms exist today, yet this number is only a fraction of 1 percent of all the species that have existed on earth since life began. More than 99 percent of all living species on earth have become extinct.

All told, then, how many species of life has evolution produced on earth since the planet's birth some 4.6 billion years ago? The staggering figure is in the vicinity of ten billion.

Now what if this picture is essentially the same for millions of other planets? What if it is actually true that in our Milky Way galaxy alone— with its hundred billion stars—this kind of species evolution has occurred? What if around every hundredth star there is a planet on which life has developed and species have proliferated? Then if we choose to stretch our imaginations beyond our own whirlpool galaxy and remind ourselves that the known universe is composed of at least a hundred billion galaxies like our Milky Way, the number of wandering biomes like Spaceship Earth is overwhelming. What about the number of species of life on all these planets? The abundance of life in our biocosmos is beyond all human comprehension.

Of all earth's diverse life-forms, how many intelligent species have evolved? We're not sure. Rudimentary intelligence is found in some higher primates, and perhaps an advanced awareness is experienced by dolphins. It could be the case that we are surrounded by a variety of other intelligences, but that our egocentric perversity has prevented our conceding their existence.

Nevertheless, one conclusion is certain: Only one species has developed culture and technology. Man's power to domesticate nature has never been challenged.

In terms of the ways of evolution, there seems to be no good reason why a biome should evolve but one intelligent species. On other planets several highly intelligent species belonging to different branches of their evolutionary tree may have evolved together and continue to coexist. In appearance they may differ greatly from one another, but in matters of intelligence they could all be similarly advanced. A multi-intelligence evolutionary pattern may be common.

The very notion that life is unique to planet earth rings in our minds today like an arrogant aristocentrism, a leftover chauvinism of the sort we recognize in other areas of our thinking and are gradually outgrowing. These inordinate claims to uniqueness and centrality—so understandable before man became informed about himself and his relationship to the natural world—sound like medieval anachronisms. Today such claims just feel wrong, and they no longer compete for intellectual credibility.

After millions of years we are at last launched on a search for our cosmic neighbors—albeit they are very distant neighbors. We can't quite hear their pets bark at night or see the smoke waft upward from their chimneys. Still, this is our last frontier, and we have embarked on a search that will change our lives.

The discovery of the existence of ETIs will begin to change man's image of himself; and the image we hold of ourselves, individually and collectively, determines what we are and how we behave.

Some futurists predict that such a discovery could be the beginning of the end of man, not because ETIs may be hostile, but because their level of existence might be so advanced that, by comparison, we humans will then appear to ourselves as little more than toddlers in the cosmic kindergarten. We would have to reclassify ourselves among the lower forms of life. That would hurt.

To those of us who are more optimistic, the discovery of extraterrestrial intelligence could be the concrete event from which a new mankind would be forged—forged partly from the new knowledge and wisdom gained, partly from the creative responses that new conditions would force from us, partly from *the confrontation with ourselves* that such an encounter would draw us into.

About 460 B.C. the Greek tragedian Aeschylus wrote that man, because of his nature, is subject to "an epistemic law" decreed by Zeus, "who taught men to think" and "has laid it down that wisdom comes alone through suffering."

An event powerful enough to force us to rethink the very foundations of our existence might succeed in producing the psychic suffering required to propel transition-man onto a path that will lead him to transcend his present condition.

In 1960 Ray Bradbury wrote, "In our time the search for extraterrestrial life will eventually change our laws, our religions, our philosophies, our arts, our recreations, as well as our sciences."

In 1973 a similar note was sounded by Carl Sagan: "In a very real sense this search for extraterrestrial intelligence is a search for a cosmic context for mankind, a search for who we are, where we have come from and what possibilities there are for our future—in a universe vaster both in extent and duration than our forefathers ever dreamed of."

Since the beginnings of awareness man has dreamed of touching the sky and talking with heavenly beings. In myth we once built a tower up which we could scale the heights to visit the gods in their council chambers. In fantasy we approached the vaulted firmament and poked our heads through the brazen dome to see what lay beyond.

Space now exerts an irresistible attraction upon us, like a magnet, drawing us outward from the confines of our Spaceship. It invites us to explore a universe that promises to be so meaningful that we will be willing to invest our future in it.

We will not again be able to perceive the cosmos as a mere matter/ energy machine. To be sure, we will always be awed by the plain physics that drives the universe. Cratered moons, planets, supernovas, neutron stars, quasars—these and a host of heavenly objects will forever hold our fascination. And, happily, we'll continue to create in ourselves the fantasies that are made of moon dust and Milky Way dreams. Still, until now the cosmic inventory contained only these marvelous manifestations of matter-in-motion. Nothing more.

Now, for us the heavens have changed. We are being drawn outward by the search for life. That makes all the difference.

THE STORY OF LIFE (to be continued . . .)

Long before the emergence of the genus Homo, hominid skywatchers must have blinked in wonderment at the stars, but such early stirrings belong to prehistory and may elude us forever. In this essay Isaac Asimov thumbs through the pages of man's recorded history and gives us a summary of speculations about life on other worlds.

Asimov is a biochemist (a Ph.D. from Columbia), a science writer (where he sticks to the facts), a science-fiction writer (where he generally sticks to future facts—he received a Hugo award in 1963 for "putting the science in science fiction"), a historian, a futurist, a Bible scholar, a textbook author . . . *etc.*: choose your label.

Asimov is the author of 172 books (174 by the time this book goes to press!). With good reason he has been called "one of our great natural resources." Some of his familiar titles are *I, Robot,* the *Foundation* trilogy, *The Hugo Winners* (editor, 3 vols.), *The Human Brain, The Human Body, Opus 100.* Most recently he has written *Eyes on the Universe* (a history of the telescope), *The Ends of the Earth* (on the earth's polar regions), and *Murder at the ABA* (a mystery novel).

What about Opus 200? "I should be able to reach Opus 200 before the end of the seventies." And Opus 300 and Opus 400? "Well, *if* I should live to be ninety, and *if* I continue to write at my present rate, I might even reach Opus 500!"

"Those who thought of other worlds invariably imagined life upon them—almost always intelligent and rather manlike life. There seems to have been an unspoken assumption that God would not waste a world."

Isaac Asimov

Of Life Beyond:
Man's Age-Old Speculations

If we are going to speculate about extraterrestrial life, we must first ask what we mean by the phrase.

We can define terrestrial life without difficulty. That would include the millions of species of plants, animals, and microorganisms that now exist on earth or have existed on earth in the past. We can then define extraterrestrial life as all forms of life that do not now exist on earth and have never existed on earth.

If that be so, then almost all human beings who have ever lived on earth, including the vast majority of those who live on earth now, firmly believed or believe in the existence of extraterrestrial life of any or all of a number of kinds.

There are, to begin with, life-forms that would fit into the natural scheme of things perfectly well, if they only existed. These would include unicorns, dragons, rocs, and similar creatures. They possess only animal intelligence, though they are more beautiful or more dreadful than the

33

animals well known to man.

Then there are various kinds of paranatural life-forms, those that follow laws of behavior radically different from those we know and that very often possess abilities far beyond human experience. Usually such life-forms are marked by human or superhuman intelligence.

There are life-forms of this sort that are more or less natural in their attitude toward humanity. There are talking animals, for instance. There are the elves, fairies, gnomes, and "little people," generally with attitudes toward humanity varying from the slyly mischievous to the mildly benevolent. There are also the various ghosts and spirits of the dead. If the readers of this article do not accept the existence of such life-forms, they should recognize themselves to be a rather inconsiderable minority of the human race even now.

Among the actually superhuman intelligences that have been widely accepted, both in the past and in the present, there are those that are malignant and that are consumed with evil intentions toward humanity—the entire tribe of ogres, afrits, demons, right down (in our own particular culture) to the Devil himself.

Finally, there are the superhuman intelligences that guide the universe and that, more often than not, are viewed as vastly benevolent; all the divine and semi-divine creatures of legend, up to (in our own culture) the angels, archangels, and God himself.

Yet all of these we can eliminate from consideration here. Whether any of them exist or not is irrelevant. When we speak of extraterrestrial life today, we are referring to those forms of life that are bound by the same laws of the universe that bind us, but that differ from the life we know only in that they occur not on earth but in some other part of the universe.

It might be more accurate to speak of extraterrestrial life as "otherworldly life" were it not for the fact that, if we conceived of life existing in dust clouds filling the vast spaces between the worlds, that too would be a form of extraterrestrial life.

If we come down, then, to extraterrestrial life in the astronomic sense, it is clear that there can be no sensible speculations concerning such things until the time when human beings understand that there are indeed worlds other than the earth.

This is by no means self-evident. In the earliest legends, the whole visible universe was conceived of as a relatively small patch of flat earth, over which was draped a solid firmament, coming down to the horizon on all sides. In the firmament the stars were mere luminous pinpoints, and the sun and moon were lamps with no other function than to light the earth and were probably not much larger than ordinary lamps. The earth, by this view, was all the habitable universe there was, and nothing could lie beyond but the paranatural abodes of gods or demons.

What upset this limited picture of the universe was the moon. All the other heavenly bodies seemed to be pure light, unchanging and eternal. The moon, however, had visible shadowings on its face that spoiled its perfection. What's more, the moon changed shape, from a perfect circle of light to half a circle, then to a crescent, then to a circle again.

Anyone studying the changing position of the moon relative to the sun during the course of these phase changes was bound to see that the lighted portion of the moon always faced the sun. From this, one could conclude that the moon was a dark body that shone only by the reflected light of the sun and that the phase changes were the inevitable consequence of the movements of moon and sun across the sky. The mere fact that the moon was intrinsically dark gave it at least one property in common with the earth, marked it with a worldlike aura, and produced the first hint of the multiplicity of worlds.

As for the dim markings on the moon, it was inevitable, given the anthropocentrism of humanity, that they would be seen as the barely made out form of a living human being on the moon. It is this sort of thing that gave rise to the legends, in our own culture, of "the man in the moon," supposedly the biblical Sabbath-breaker of Numbers 15:32-35. The man-in-the-moon legends may represent the first speculation concerning extra-terrestrial life in the modern sense.

Yet though the moon was a dark world like the earth, it could, of course, be pictured as no larger than it seemed to be; so that the man in the moon filled the world he lived on.

About 130 B.C., however, a Greek astronomer, Hipparchus of Nicaea, calculated the distance from the earth to the moon accurately for the first time, and found it to be thirty times the earth's diameter. Combining this with the best Greek estimate of the earth's diameter, the moon's distance from earth turned out to be 240,000 miles. To appear as large as it does in the sky from that distance, the moon must be some two thousand miles across.

This was the first definite indication in human history that at least one other world existed in the universe, a real world with dimensions that were a respectable fraction of the earth's size.

If that were so, what of the other planets, whose distances could not be determined by the ancient astronomers but that were known to be farther off than the moon? Might they too be worlds like the earth and the moon?

The conclusion was a fair one, and the lively imagination of writers could seize upon that thought. In the second century B.C. the Syrian writer Lucian of Samosata wrote the first interplanetary romance that we know of. He tells of a ship that was carried up to the moon by a waterspout. He describes the intelligent beings living on the moon and tells of the war they are conducting against the intelligent beings of the sun over their conflicting ambitions to colonize the world of Venus.

What seemed true, indirectly, from Hipparchus's measurement of the distance to the moon was backed by more direct evidence in 1609, when the Italian scientist Galileo Galilei turned a telescope on the heavens for the first time. When he looked at the moon he saw a world on which there were mountains and craters. He also saw flat regions he called seas. The moon was visibly a world like earth, in these respects at least.

When Galileo turned his telescope on the planets, he saw them magnified into small globes—small, it was obvious, only because of distance. The moon and the planets were worlds, beyond dispute, and this fact gave a new impetus to dreams of extraterrestrial life.

In 1638 an English clergyman, Francis Godwin, wrote a story called *Man in the Moone*. Godwin's hero flew to the moon in a chariot hitched to great geese that were supposed to migrate to the moon regularly. Godwin described the moon as a world much like the earth, but better. This book was the first of the line of modern interplanetary romances, a line that has continued to flourish to this day and that still forms a major theme in the fabric of contemporary science fiction.

In general those who thought of other worlds invariably imagined life upon them—almost always intelligent and rather manlike life. There seems to have been an unspoken assumption that God would not waste a world. If worlds existed, their only purpose had to be to bear manlike beings.

Nor was it only romancers who thought so. Consider the German-English astronomer William Herschel, who in the decades immediately preceding and succeeding 1800 was the foremost astronomer in the world. Herschel could not even bear to spare the sun. He suggested that the sunspots were holes in the flaming atmosphere of the sun, that through them the dark and cool surface of the sun's globe itself could be seen, and that

this dark and cool surface might be inhabited.

Even as late as 1835 the *New York Sun* was able to run a series of hoax stories about a moon on which there were earthlike conditions and intelligent life, and found (perhaps to its surprise) that the general public believed them. The *Sun*'s circulation boomed until the hoax was exposed.

Yet it was not long after Galileo's initial observations of the moon by telescope that it became rather obvious that the moon's seas were not of water but of flat, dry land, and there were no signs of either water or air on our satellite. There were no clouds on the moon, no twilight, no change of any kind. It was an airless, waterless world.

Of course, it did not follow, inevitably, from the fact that no air or water could be detected on the moon that it was without life. Yet, if there were life on the moon, it would have to be radically different from ours, and there was no evidence that such "other life" existed. Among astronomers it grew steadily more common to view the moon as lifeless. It was thus borne in upon mankind, for the first time in history, that it was possible to have a dead and barren world.

The first interplanetary romance that tried to take into account the state of scientific knowledge of the time (and was therefore perhaps the first real science-fiction story) was written by the German astronomer Johannes Kepler. It was published posthumously in 1634. In this tale, *Somnium,* the hero is transported to the moon in a dream. There he found a day and night that were each two weeks long (something that even primitive stargazers knew, but a fact that romancers almost invariably ignored). However, Kepler could not abandon life. He postulated strange animals and plants that grew rapidly during the long day and died at nightfall.

In 1643 another blow at romance was struck by the Italian physicist Evangelista Torricelli. In that year he invented the barometer and showed the pressure of the atmosphere to be equal to that of a column of mercury thirty inches high.

If the atmosphere exerted only that great a pressure, it would have to extend upward no more than five miles, assuming it to be the same density all the way up. In actual fact the density of the air decreases rapidly with height, so that the atmosphere extends much higher than five miles, but it rapidly grows too thin to support life.

The universal assumption of earlier times that air extended indefinitely upward (so that men could travel to the moon in chariots hitched to flying geese) was destroyed. It became clear that the various worlds of the universe were separated by vast stretches of vacuum that acted as an insulating layer as far as life was concerned. If there were life on other worlds, reaching them or having them reach us would be a matter of enormous difficulty.

Oddly enough, the first suggestion of the one possible means of reaching the moon through the vacuum beyond the earth's atmosphere came not from a man of science but from a science-fiction writer—none other than Cyrano de Bergerac. Cyrano, the long-nosed duelist, really existed, really had a long nose, really fought duels, and was also a clever writer. In 1650 he published a book called *Voyage to the Moon*, which treated the moon as a world inhabited by intelligent beings. In the course of the book, he suggested several methods of reaching the moon, and one was to tie rockets to a chariot, light them, and zoom off.

To this day the rocket engine is the one propulsive method known to man by which it is possible to cross a stretch of vacuum and steer a course through space. It is the method by which mankind finally (three centuries after Cyrano's tale) reached the moon.

By the nineteenth century the solar system was found to contain many more worlds than had been known in the days before the telescope. A seventh planet, Uranus, had been discovered in 1781, and an eighth, Neptune, in 1843. It was known that the outer planets were far larger than earth, and that each had a train of satellites that, taken together, represented some dozen and a half additional worlds of varying size. There were numerous small asteroids circling the sun in the space between the orbits of Mars and Jupiter.

However, nineteenth-century astronomy had grown quite sophisticated, and as more and more was learned about these various worlds, the outlook for other life grew more and more gloomy. In the 1860s the Scottish mathematician James Clerk Maxwell worked out the molecular interpretation of the gas laws, and it came to be understood why the moon was airless. Its gravitational field was insufficiently intense to hold the rapidly moving molecules of gases to its surface. It would seem from this that any world as small as the moon or smaller would have no atmosphere or, at best, one too thin to support our kind of life. All the satellites and asteroids were therefore airless, or nearly so.

Mercury was larger than the moon, but it was also closer to the sun and therefore much hotter; that meant it, too, was airless. The larger outer planets did have atmospheres, but those planets were so large that their gravitational fields were intense enough to collect and maintain atmospheres of such enormous pressure as to make their environments utterly unearthlike.

Again, we cannot utterly eliminate the chance that there are strange life-forms capable of enduring conditions we would consider intolerable, but there is not even the faintest evidence of the existence of life of a nature wildly different from our own.

Some have indeed speculated on the kinds of chemistry that might form the basis of life on worlds much colder or warmer than our own, or with radically different atmospheres and oceans. There have been thoughts of ammonia's playing the role of water on cold planets, of silicones or sulfur doing so on hot planets, and of complex silicates building themselves into life-forms on liquid-free planets. Such speculations, barren of any observed evidence at all, lead nowhere unfortunately, and, on the whole, astronomers have found it more useful to restrict themselves to the possibility of extraterrestrial life similar in basic chemistry to our own.

The question was, then, whether any world did or could possess the type of environment to which life similar to the life-forms on earth could possibly adapt itself.

By the second half of the nineteenth century, the possibilities of this, among the known worlds at least, had been reduced to two and two only, other than earth. These were the planets Venus and Mars.

Venus is almost as large as earth and is closer to the sun. It might be warmer, but perhaps not too warm for life. It obviously has an atmosphere, and on it is borne a permanent layer of clouds that would protect the surface from the too-ardent embrace of the sun. It was easy, then, to speculate that Venus was a watery, primitive world that might even have a planet-girdling ocean. However, since there was no way of peering through the eternal cloud layer at the surface below, there seemed no way of advancing beyond that speculation.

As for Mars, it is smaller than earth and has a thinner atmosphere, yet perhaps not too thin for life. It is farther from the sun than is the earth, therefore colder, but perhaps it is not too cold for life. The axis of rotation of Mars is tipped to its plane of revolution about the sun by almost exactly the amount earth's axis is tipped. This means that Mars has earthlike seasons (though colder). It has icecaps at both poles that expand and contract with those seasons, and color changes that seem to represent advancing and retreating vegetation.

Mars is unique in another respect. Of the worlds beyond the moon, only Mars is close enough and has an atmosphere clear enough to allow its surface to be mapped. Naturally the best map would be produced when Mars makes a particularly close approach to earth.

Every two years or so, earth passes Mars in their two orbital motions about the sun. These points of passing ("oppositions") come at different places in the planetary orbits, orbits that are closer to each other in some places than in others. In 1877 earth was in opposition to Mars at points in their orbits that were nearly at minimum separation. Mars could then be seen with the kind of clarity that comes only three times in a century, and many astronomers planned to study it then.

One of those who studied Mars at the 1877 opposition was the Italian astronomer Giovanni Virginio Schiaparelli. He produced the first modern map of Mars, one that persisted essentially unchanged until the development of techniques for observing Mars that proved far more powerful than the telescope.

In drawing his map Schiaparelli noted dark areas, which he considered to be bodies of water, and light areas, which he considered to be land. Connecting the dark areas, Schiaparelli saw a few markings that were rather straight and thin. He thought these were narrow connecting waterways, rather like the various straits and channels on earth. Indeed he called them *canali,* which is the Italian equivalent of "channels" and means "narrow waterways."

Unfortunately *canali* was translated into English as "canals," which are, of course, artificial waterways. The notion arose and grew stronger that Mars, with a gravitational field at its surface only two-fifths of that at earth's surface, was gradually losing its water; that an ancient, dying civilization on that planet was intent on preserving as long as possible what water still existed and making as efficient use of it as possible. The canals, it was thought, were an advanced engineering plan to keep the planet irrigated. For the first time it seemed that science had produced direct evidence not only for life on another world but for *intelligent* life and a highly developed civilization.

In 1894 the American astronomer Percival Lowell established an observatory in Arizona. For years he studied Mars closely, through the clear air of the southwestern skies, and drew maps showing an intricate lacework of canals. Eventually he plotted five hundred of them on his maps. He also located the "oases" at which they met, recorded the fashion in which the canals seemed to double at times, and noted in detail the seasonal changes that seemed to mark the ebb and flow of agriculture. Lowell, who in 1908 published a book entitled *Mars as the Abode of Life*, was the foremost proponent of the theory of life on Mars.

Even more influential than Lowell in convincing the public that there was intelligent life on Mars was the English science-fiction writer H. G. Wells, who responded to the growing popularity of the Schiaparelli/Lowell view by publishing *The War of the Worlds* in 1898. In this book Wells told of Martians who, despairing of being able to maintain their dying world, launched an invasion of earth. Their superior technology made it possible for them to overwhelm the inhabitants of earth, but the Martians were defeated by physiology, for their bodies could not resist the onslaught of earth's decay bacteria. This was the first story to deal with interplanetary warfare, and it introduced the first note of fear into speculation about extraterrestrial life.

The belief that there was a civilization on Mars remained popular with the public well into the mid-twentieth century. In 1938 a realistic version of *The War of the Worlds* was performed on American radio by Orson Welles. Martian ships were reported to be landing in New Jersey, which created panic among parts of the population there.

Astronomers, however, did not by any means all agree on the existence of the Martian canals. Many excellent observers did not succeed in seeing canals on Mars. The feeling arose among some of them that the canals were an optical illusion, that there were irregular blotches on the Martian surface that, individually, were just below the limits of vision but that, taken together, were interpreted by straining eyes as straight lines.

There was no way of deciding between those who supported canals and those who denied them as long as only telescopic views were possible; but throughout the twentieth century, as more and more was learned about Mars, the less hospitable its world seemed to be. Mars was colder and dryer than had been thought. Its atmosphere was thinner and contained no oxygen. Even the icecaps might be frozen carbon dioxide rather than water.

Finally, in 1965 the Mars probe, Mariner 4, took photographs of the planet from a distance of six thousand miles above its surface. The photographs showed many craters but no canals. Still more sophisticated probes since then have succeeded in mapping the entire Martian surface in detail. There are giant volcanoes and canyons and many other fascinating features, but no canals. What Lowell saw were indeed optical illusions.

It may still be that there are simple forms of life on Mars, bacterialike or lichenlike forms. It may be that Mars suffers alternations of climate and that we are now viewing it in its "ice age" period. Under other conditions its environment may be much milder and more hospitable to life. A decision about that will perhaps have to wait for a manned landing on Mars, but a dispassionate assessment of the situation must make it appear that the chance of finding life on Mars is rather low, while that of finding *intelligent* life is virtually nil.

In 1962, meanwhile, the Venus probe, Mariner II, confirmed what astronomers had been suspecting for several years—that Venus was extraordinarily hot. Its thick atmosphere, a hundred times as dense as our own, was almost entirely carbon dioxide, and such an atmosphere trapped heat (the "greenhouse effect") and raised Venus's surface temperature to over 400° C. Far from being a waterlogged planet, Venus is incredibly hot and bone dry. Life as we know it cannot possibly exist on Venus.

Finally, in 1969 human beings stood upon the soil of another world—the moon—for the first time and found it to be, as expected, completely lifeless.

So it would seem that at the moment of this writing there is only one world in the solar system, outside earth itself, on which there may be life. That world is Mars, and even there any life that exists will probably be very simple—and the chance that any life at all exists is small.

We have no choice but to admit that, excluding earth, the odds are rather heavily in favor of our solar system's being dead.

There is, however, a vast universe outside the solar system. What of that?

Beyond the solar system are the stars, but they were for a long time discounted. In early times, when the moon and sun seemed to be worlds and when it was already possible to assume the planets might also be worlds, the stars were still viewed as merely decorative markings on the solid curve of the sky.

The first person we know of who thought of stars as worlds was a German cardinal, Nicholas of Cusa, who in 1440 published some remarkably modern-sounding notions of the universe. He held that space was infinite and that the stars were other suns. Since it seemed unnatural to him that all of those suns would be wasted, he assumed that each had its family of planets circling it and that those planets were inhabited.

The Church was secure in 1440, and Nicholas of Cusa did not get into trouble over his views. A century and a half later, the Italian philosopher

Giordano Bruno vehemently upheld views similar to those of Nicholas, but this was at a time when religious disputes were racking Europe. In 1600 Bruno was burned at the stake for his various heresies.

The coming of the telescope and the revelation that the outer planets were hundreds of millions of miles distant made it clear that the stars, which had to be farther off still, must be extraordinarily luminous to be visible at all. More and more astronomers began to think of them as distant suns.

In 1718 the English astronomer Edmund Halley reported that some of the brighter (and, therefore, possibly nearer) stars, such as Sirius, Procyon, and Arcturus, had shifted position since ancient times. This demonstrated that the stars were not fixed in the sky but were freely moving, like individual bees in a swarm. If they appeared motionless, it was because their enormous distances made their motions seem very slow. Halley's discovery put a final end to any notion that the stars might be anything other than suns.

It is not surprising that when the French satirist Voltaire, in a story called *Micromegas,* published in 1752, described two extraterrestrial visitors, one of them came from Sirius.

It was not till 1838, however, that Friedrich Wilhelm Bessel, a German astronomer, first measured the distance to a star. The star he chose to study was a rather dim binary star (consisting of two stars close together and circling each other) named 61 Cygni. He chose it because it moved against the background of the stars more rapidly than any other star was known to, thus indicating that it might be unusually close. Checking its slight shift in position as the earth moved around the sun, Bessel estimated 61 Cygni to be some sixty million million miles away.

Although light travels at the speed of 186,000 miles per second, it takes light eleven years to cover this great distance. The star 61 Cygni is therefore eleven "light-years" away. Even the nearest of all stars proved to be a little over four light-years away.

In the century that followed, the known size of the universe constantly increased. The stars we see in the sky are all part of a system called the galaxy, and the Milky Way is what we see when we look through the long axis of the galaxy—endless numbers of faint stars. Actually, it is estimated there may be 135 billion stars in the galaxy, and there may be as many as 100 billion other galaxies distributed through space. The total number of stars in the universe is quite beyond comprehension, but so are the distances that separate them.

Nor can we conquer those distances by building up equally vast velocities for our spaceships. In 1905 the German scientist Albert Einstein demonstrated that the speed of light in a vacuum was the fastest speed that massive objects could ever attain. Any probe travelling to even the nearest star at even the greatest conceivable speed would have to take 4.3 years for the journey.

It seemed discouraging to have to speculate about extraterrestrial life in any worlds located at the vast distances of the stars. As long as it seemed that there might be extraterrestrial life in the solar system, then, astronomers wasted little time speculating about the stars.

One route by which stars entered into speculations dealing with the planets involved the question of the origin of planetary systems.

In 1798 the French astronomer Pierre Simon de Laplace had suggested that the solar system was originally a large cloud of gas and dust (a "nebula") that whirled majestically and slowly condensed into the sun. As it did so it gave off small portions of itself that formed the planets.

Over the course of the nineteenth century, however, Laplace's "nebular hypothesis" crumbled under the weight of various objections. It was hard to see, for instance, how enough of the whirling property ("angular momentum") of the cloud could be concentrated into the small portions that formed the planets. Some 98 percent of the angular momentum of the solar system is concentrated in the planets, which, taken together, are only one-thousandth as massive as the sun.

In 1917, therefore, the English astronomer James Hopwood Jeans suggested an alternative explanation. He proposed that the sun and some other star had had a near collision and that the gravitational pull of each had served to extract matter from the other. This extracted matter was given a strong spin by the gravitational fields of the stars as they moved apart. By this theory, which was widely accepted for over a quarter of a century, planetary systems arose only when two stars passed close to each other. Considering the distances that separate the stars and the speed with which they move, such a near collision must be extremely rare. In the entire history of the galaxy it might have happened only once.

By Jeans's theory, then, there would possibly be only two planetary systems in our galaxy—our own and that of the star that passed close to the sun. And it might be that in these two planetary systems our own planet is the only one that bears life. The vision to which Jeans's theory gave rise, then, was of a glorious universe, utterly free of life except on the earth and, at most, on one or two other such planets per galaxy.

Jeans's theory had shortcomings, too, and these grew to seem more serious as astronomers' knowledge of the inner structure of stars increased. In 1944, the German astronomer Carl Friedrich von Weizsäcker suggested a return to the theory of Laplace at a more sophisticated level. Turbulence of the original dust and gas was taken into account, and in later modifications of the theory electromagnetic forces as well as gravitational ones were considered. If Weizsäcker's theory is correct, then planetary development is a normal step in the birth of a star, and nearly every star ought to have planets circling it.

Even as Weizsäcker was proposing his theory, observational evidence was supporting it. In 1943 the Dutch-American astronomer Peter Van de Kamp reported small irregularities in the movements of one of the two stars of 61 Cygni. These small irregularities, it could be deduced, were caused by the gravitational pull of a large planet eight times the mass of Jupiter. In 1960 a planet of similar size was shown by Van de Kamp to be circling the small star Lalande 21185, and in 1963 a smaller planet (possibly two) seemed to be producing irregularities in the motion of Barnard's star.

Stars are so distant that it takes a particular set of conditions to make their planets detectable. The stars must be very small and the planets especially large, so that the gravitational effects of the latter on the former are considerable. As it happens, Barnard's star is the second closest star to earth, Lalande 21185, the third closest, and 61 Cygni, the twelfth closest. That three planetary systems should be detected in our immediate neighborhood, even though the requirements for detection are stringent, is extremely unlikely unless planetary systems are common indeed.

At the present time, therefore, astronomers are generally convinced that planets are common phenomena and that most or all of the stars in the universe possess some sort of planetary system and that there are therefore innumerable worlds on which life might conceivably develop.

By the middle of the twentieth century, of course, scientists were no longer ready to assume that any world would have life as a matter of course. There was no longer the feeling that a world without life was a world wasted; experience with the solar system itself had revealed how easy it was for a planet to present an environment so basically different from that of earth that life as we know it was not in the least likely to exist.

It was legitimate to wonder, then, if worlds might not remain barren even though they seem, to our eyes, to be potentially hospitable to life.

Perhaps even though planets are a common and even inevitable phenomenon, life is not. Life might originate only by an incredibly rare chance; and the universe, though it might be studded with likely planets, might be empty of life except for that on our own world.

A definite impression that life might not be the result of a rare accident came about through a series of experiments that began in 1952 with the work of the American chemist Stanley Lloyd Miller. Beginning with a sterile mixture of simple compounds of the type that might have existed on the primordial earth—hydrogen, methane, ammonia, and water—Miller added energy in the form of an electric discharge, and in a week's time he found that more complicated compounds had been formed, including two of the amino acids that form the building blocks of protein molecules. Later experiments by others made it quite clear that simple compounds plus energy yielded complex compounds that were invariably of types that seemed to point in the direction of life.

Nor was this apparent only in the laboratory. Beginning in 1968 signs of atom combinations were discovered in the dust clouds found in interstellar space. Through their absorption of specific radio wavelengths, more than a score of compounds revealed themselves to astronomers. Several of these, too, pointed in the direction of life.

It seems quite certain, then, that life is the result of a natural tendency, an obvious route taken by compounds under the whip of energy toward a steadily growing complexity. We might suppose that had a different route been the natural and obvious one, life would have begun in another fashion, based on another kind of chemistry—but one route to complexity was *the* route and it gave rise to *our* kind of life.

Since life would seem not to be the result of a rare accident, and since life, approximately as we know it, would seem to be on the natural line of development for any planet similar in mass, temperature, and chemistry to our own, some astronomers have begun to calculate probabilities. They estimate the number of stars per galaxy that may be enough like our sun to give their planets an environment like that of our own solar system, then how many of these planets may be enough like earth to develop our kind of life, then how many of the life-bearing planets may develop civilizations, and so on. One of the most assiduous and persuasive of these calculators is the American astronomer Carl Sagan, who is perhaps the outstanding "exobiologist" (one who studies the possible nature of extraterrestrial life) in the world.

The American astronomer Stephen H. Dole, in his book *Habitable*

Planets for Man, published in 1963, took into account all the factors of planetary systems that could be deduced (sometimes shakily) from the one system we know—our own. He suggested that our galaxy might have as many as 640 million earthlike, life-bearing planets and, presumably, that there may be as many as that in each of the other galaxies, on the average. He also concluded that there is a 50 percent chance of finding an earthlike, life-bearing planet within twenty-two light-years of earth.

To be sure, since 1963 added knowledge concerning the universe has made it seem increasingly to be a desperately violent place. There are quasars that are far smaller than galaxies, yet far brighter, neutron stars that have arisen out of gigantic stellar explosions, black holes that inexorably swallow up all neighboring matter and give back nothing at all. It may be that the nucleus of every galaxy is the scene of violent events that produce an environment inimical to life. It may be that we might expect life to exist only in the sparsely distributed stars in the quiet suburbia of a galaxy's spiral arms.

Only 10 percent of the stars in a galaxy are located in its spiral arms, so that the chance of life in the universe might be viewed as falling dramatically to only 10 percent of what might have been thought a little over a decade ago. Even so, Dole's calculations might yield sixty-four million earthlike, life-bearing planets in the spiral arms alone, and that is not exactly a small number. And since our own solar system is located in one of the spiral arms of our galaxy, projections about our near neighbors, within, say, ten thousand light-years, would remain unaffected by this view.

The sort of astronomical thinking that since World War II has made it seem that both planets and life are overwhelmingly common in the universe has, of course, had its effect on popular thinking. Quite aside from science-fiction stories, which have frequently depended on a great number of planets and a great deal of life to thicken their plots, new cults involving extraterrestrial life have arisen.

There is, for instance, the flying-saucer cult, which grew up immediately after World War II. Its adherents assume that spaceships from other worlds are regularly observing earth. As it becomes less and less possible for anyone, even cultists, to suppose that there are space travelers from any of the worlds in the solar system, there has been a tendency to assume they come from worlds circling other stars.

In the early 1970s came the cult centering on Erich von Däniken's *Chariots of the Gods*, which supposes that travelers from outer space landed on earth in prehistoric times, taught mankind its technologies, are responsible for some of the artifacts of primitive cultures, and may even have contributed alien genes to earth's native life.

Scientists do not take these cults in the least seriously, but neither do they entirely discount the possibility of communication with extraterrestrial life. The difficulties of such communication are, however, perhaps too easily dismissed by the cultists.

First, the distances between stars remain a strongly insulating phenomenon. Even if we suppose with Dole that there is a 50 percent chance of finding an earthlike, life-bearing planet within twenty-two light-years, such a distance is still not exactly close. It may be tiny compared with the size of the universe, or even with the size of the galaxy; but, on the other hand, it is thirty-two thousand times the distance from here to Pluto, our system's farthest planet.

To be sure, we might conceive of ways to get around the barrier of the speed of light as a maximum. There is no evidence at all, however, that any conceivable scheme for faster-than-light speed exists or can exist. For the moment, then, we must assume that the speed of light remains the limit.

Even so, if we are willing to take enough time to begin encroaching on eternity, we may nevertheless attempt to communicate with extraterrestrial life. On March 2, 1972, the Jupiter probe, Pioneer 10, left earth. It passed Jupiter on December 3, and then moved beyond; in 1984 it will leave the solar system altogether.

It carries a 6- by 9-inch gold-covered aluminum slab, designed by Sagan and his fellow-astronomer Frank Donald Drake and drawn by Linda Sagan. The slab carries information about the nature of the solar system from which it originated and about the living beings who made it, but it may travel millions of years before it is picked up. And . . . it may *never* be picked up.

To send a material object is, however, not a very satisfactory method of attacking the problem of communicating with extraterrestrial life. When astronomers think of communicating with other planetary systems, they usually think in terms of massless signals, such as light or radio waves that travel at the speed of light, thousands of times faster than Pioneer 10.

Thus, in the early 1960s, Drake supervised a brief attempt to detect radio signals from the direction of several stars that were thought to be sunlike enough to have some chance of possessing life-bearing planets and close enough to have some chance of allowing radio signals to reach us in detectable intensities. Nothing was detected, but the thought remains. In

1969, when the English astronomer Anthony Hewish picked up for the first time rapid pulses of radio waves of a type never before detected, the first exciting thought was that it was a message of intelligent origin. It was referred to at first as an "LGM" phenomenon, the initials standing for "little green men." It turned out, however, that the signals were received from rapidly rotating neutron stars—fascinating in themselves, but not nearly as fascinating as the discovery of extraterrestrial life would have been.

Another uncertainty, perhaps the most serious, remains. Even though planets may be a universal accompaniment of stars, and even though life may be a universal accompaniment of earthlike planets, how sure can we be that the development of intelligence in the course of evolution is inevitable?

On earth, life existed for some three billion years before any single species developed a large brain in an environment that made the development of a technological civilization possible. But was such a development inevitable? Might not life on earth have continued indefinitely without developing intelligence and a civilization? Is it possible that intelligence, unlike life itself, is the product of an extremely rare combination of events?

And even if intelligence does develop frequently in different worlds, might it not be that once it evolves it may (judging from our own sad example) soon thereafter destroy itself?

If this were so, then, for all we can at present say, the dream of communication with life-forms in other planetary systems may be hopeless. It may be that out there are only civilizations that have not yet come to be or that have come and quickly ceased to be; it may be that we can talk only to ourselves during the brief interval before we ourselves, as a technological civilization, cease to be.

And yet the sanguine Sagan speculates in reverse and wonders whether there might not be a widespread Empire of the Stars, in which intelligent beings communicate and interact by way of scientific techniques undreamed of by us.

He speculates, for instance, that advanced science may make it possible to use black holes to link together a space-vaulting civilization; that devices may be used to plumb the depths of black holes where the laws of nature (as we know them) lose their validity and where instantaneous travel across vast stretches of space, and even forward and backward in time, may be possible.

This sounds rather less plausible, if anything, than do flying saucers and ancient spacemen, but what if Sagan's dreams are true?

Imagine, then, how incredibly exciting it would be to touch, at last, even the outermost fringes of this godlike knowledge of life beyond.

Any feeling of optimism about finding life in our solar system?

"Not really. But it would be exciting, wouldn't it!"

What about contact with extrasolar ETIs?

"It would be of ultimate significance to discover even one civilization that has endured for tens of thousands of years. At least if those fellows made it, perhaps there is hope for us!"

For the present anthology George Abell provides a general summary of the scientific framework within which biocosmic research and speculation is currently taking place.

George Abell is, among many other things, Professor of Astronomy at the University of California, Los Angeles. He is an alumnus of Cal Tech and has periodically served as consultant to the Jet Propulsion Laboratory and other movers of America's space program. He has collaborated on numerous research projects in astronomy at the Hale (Mt. Wilson and Mt. Palomar) observatories.

Dr. Abell is the author of a widely used textbook in astronomy, *Exploration of the Universe*, and a shortened edition entitled *Realm of the Universe*.

"Unless the processes by which amino acids and nucleo-tides become living organisms are so extraordinarily un-likely as to be extreme long shots, it would be surprising if life were not abundant in the galaxy. Our problem is to find it."

George Abell

The Search for Life Beyond Earth: A Scientific Update

THE BEGINNING OF THE UNIVERSE

According to the current popular theories of cosmology, the universe, as we know it, began sometime between twelve and twenty-five billion years ago. All matter was compressed to an enormous density and was at a temperature of some ten billion degrees centigrade. Ordinary atoms did not yet exist, only certain subatomic particles—protons, neutrons, electrons, positrons, and neutrinos. Explosively, the great mass began expanding, and it is still expanding today. During the first few hours, as the matter thinned out, it converted itself to atoms of hydrogen and helium (the helium making up about 20 percent of the mass), but for the first million years or so the universe remained a great opaque, primeval fireball, rather like the fireball in a nuclear-bomb explosion, or like the interior of a star.

We must not think of any particular site for the explosion that began the expansion of the universe; we (or at least the particles that eventually became the atoms of our bodies) were inside of it then, and still are—it is

53

all around us. It may well have been infinite in size and mass, even in those early moments of the universe, in which case it is expanding to still greater infinities today. Where the matter came from in the first place, and how it got into its hot, dense condition are questions that, at present, lie in the realm of speculation. Perhaps it resulted from a previous collapse of the universe. It is not yet even ruled out that the universe may be alternately collapsing and expanding with a period of cosmic pulsation of many thousands of millions of years.

At any rate, as the universe thinned out and cooled, its matter somehow collected into great inhomogeneous regions that formed into clusters of galaxies, each galaxy a system of billions of individual stars. Our theories of galaxy formation are still very incomplete, but we are quite sure that the oldest galaxies that we observe today were formed somewhere between ten and fifteen billion years ago.

Among them was our own galaxy, an assemblage of more than a hundred billion stars in the shape of a great wheel about 100 thousand light-years across. (A light-year, the distance light travels in a year, is nearly six trillion miles.) In addition to stars, our galaxy contains vast tenuous clouds of very sparse gas and widely scattered microscopic particles of dust spread out through much of the near vacuum of interstellar space. As stars age, burning up their internal supply of fuel for the nuclear energy that keeps them shining, they evolve. Many, late in their lives, eject much of their own material back into space, and some do this in powerful supernova explosions. But that ejected matter has been altered by the nuclear reactions that gave the stars energy in their lifetimes. Thus stars that were formed of, say, pure hydrogen and helium may eject into the interstellar medium hydrogen and helium that is "contaminated" with carbon, nitrogen, oxygen, iron, and the various other chemical elements. During billions of years, therefore, the interstellar gas and dust gradually become richer in those kinds of heavier elements that make up the bulk of the earth and other planets, and our own bodies.

THE BEGINNING OF THE EARTH

New stars continually form from the interstellar clouds. Rather recently in the history of the galaxy, just over 4.6 billion years ago, a particular cloud of interstellar material condensed; it is called the *solar nebula*. As the solar nebula contracted, it rotated faster and faster to conserve its angular momentum. Its rotation caused it to flatten into the disk surrounding a central "hub." The hub became our sun. The matter in the disk partly condensed into grains of rocky and icy matter. The grains gradually accreted each other and grew into the planets and other bodies in the solar system today. Far from the sun, where the temperature was low, large planets formed. They contained a lot of the icy material, and even attracted some of the gaseous matter from the disk as well. In the warmer, inner regions of the solar system, however, ices could not condense, and the planets that formed there—Mercury, Venus, Earth, and Mars—built up entirely by accretion of metallic and rocky particles.

The atmospheres of the earth and other inner planets have come from subsurface rocks. Heat produced by the decay of radioactive elements causes some of these rocks to partially decompose and release gases they have held in chemical combination. These gases *outgas* to the surface. On the earth, at least, the major method of outgassing is by vulcanism; the process is still continuing today. The most important gases released from the earth's crust (and probably from the crusts of the other inner planets as well) are water, carbon dioxide, and nitrogen. The water, of course, has condensed to form our existing oceans. The carbon dioxide, active chemically, has recombined with surface rocks, has dissolved in the oceans, and more recently some of it has found its way into coral formations. The nitrogen, relatively inert chemically, remains the principal constituent of the present atmosphere, although traces of water vapor and carbon dioxide are also present at any time. Most of the argon, which now makes up about 1 percent of our atmosphere, came from the radioactive decay of potassium-40.

Oxygen, which comprises 21 percent of our air, had a different origin. It has gradually accumulated, mostly over the past 600 million years, by photosynthesis in live vegetation. Atmospheric oxygen, vital to our life, was itself produced by life on the earth. On no other planet have we found appreciable amounts of free oxygen.

THE BEGINNING OF LIFE

When and how did life begin on the earth? The earliest fossils identified are of bacteria and blue-green algae, found in sedimentary rocks from 2.7 to 3.5 billion years old. Most experts believe that the primordial atmosphere contained appreciable amounts of hydrogen and other gases which outgassed from the subsurface of the earth. It is doubtful that much hydrogen could have been gravitationally attracted to the forming earth from the original solar nebula, but hydrogen was certainly present in chemical combinations in rocks. Some, for example, could have been released by the decomposition of water in the rusting of iron. At any rate, with hydrogen present, methane (CH_4) and ammonia (NH_3) would easily form, giving the young earth a *reducing* (rather than its present *oxydizing*) atmosphere.

Hydrogen, methane, ammonia, and water, mixed together, make up what the biologists call a "soup." Laboratory experiments show that when such a soup is subjected to electrical discharge (as in lightning) or ultra-violet radiation there results a large yield of amino acids and the sugars and nitrogenous bases that are the constituents of nucleotides. Amino acids and nucleotides, in turn, are the building blocks of proteins and nucleic acids—the master molecules of all living organisms. Our knowledge has not yet bridged the gap from these prebiological organic compounds to the simplest organisms, but at least we can understand how the start of the organic chemistry necessary for life can have occurred on earth.

Only carbon has the chemical versatility to provide the basis for complex living organisms, but carbon is abundant in the universe, and there is every reason to expect similar processes that took place on earth to occur elsewhere. Indeed, amino acids (but, I emphasize, nonbiological) have been found in meteorites, and organic compounds of up to nine atoms have even been detected in the gases of interstellar space. Cornell astronomer Carl Sagan hypothesizes that the red-brownish colors in the clouds of Jupiter and Saturn could well be due to the kinds of polymers of long hydrocarbon chains that have been produced in laboratory experiments with mixtures of gases known to be present on those planets.

We do not at present have any direct evidence for life anywhere except on the earth, but the formation of prebiological organic molecules would seem inevitable under conditions that must be very common in the universe. Unless the processes by which amino acids and nucleotides become living organisms are so extraordinarily unlikely as to be extreme long shots, it would be surprising if life were not abundant in the galaxy. Our problem is to find it.

LIFE ON OTHER PLANETS

We generally think of planets as the most likely abodes of life (but for an imaginative alternate possibility, read Fred Hoyle's *Black Cloud*). We cannot positively rule out simple organisms even in such a seemingly hostile environment as the atmosphere of Jupiter, but only Mars among the other planets is expected to have a realistic chance for life. Even on Mars, it is not very likely. The lack of oxygen rules out widespread photosynthesis by plants. The thin air and lack of water in the liquid state minimize molecular mobility, making organic chemistry more difficult. Still, Mars does have some frozen water as polar caps (which are mostly solid carbon dioxide), and may have subsurface water as permafrost. The surface of the planet is also traversed by branching and meandering channels that most geologists think are probably dry riverbeds. If so, Mars has had running water in past ages, and may now be in a glacial period, like one of the ice ages on earth. If liquid water has existed there for long enough periods at a time, perhaps organisms developed on Mars.

The possibility is so exciting that the United States has two Viking spacecraft on the way to the planet. Each will land an instrument package on the surface during the summer of 1976.* Although sizable plants or animals are not expected, each lander nevertheless will carry a television camera to look and see. Other experiments will scoop up Martian soil and analyze it in various ways for evidence of microorganisms, either living or dead. If the experiments work properly, and if there is as much life at either of the landing sites as there is in the middle of the Sahara Desert, and if such life is based even remotely on the same chemistry as terrestrial life, the Viking landers should detect it. Failure to detect life on Mars will not prove it is absent there, and will, of course, say nothing about life elsewhere in the universe. But positive detection of life on Mars would have, as we shall see, profound influence on our estimates of the probability of intelligent life elsewhere in the galaxy.

Many other stars must have planets, but one as insignificant in size as the earth could not be observed with our instruments. The only kind of planet beyond the solar system that we have a likely chance to detect would be a very large one revolving about a nearby star that is less massive than the sun. Such a planet might reveal itself by its very slight gravitational influence on that star. There are several nearby stars that are suspected of showing small periodic motions, as if each were in mutual revolution with a body somewhat larger than Jupiter; no such discovery, however, has been confirmed.

Nevertheless, we can make some educated guesses that enable us to

*This was written in early 1976. By the time it is read the success of the Viking mission will be history.

put some limits on the number of civilizations that may have developed in our galaxy. These limits, we shall see, are very broad, but at least they do give us some insight into the kind of procedures that might prove fruitful in our search for extraterrestrial life and intelligence.

ESTIMATES OF THE NUMBER OF EXTRATERRESTRIAL CIVILIZATIONS

Following the approach of Frank Drake, and later Carl Sagan and I. S. Shklovskii, let us consider the factors on which the number of communicative civilizations now extant in our galaxy depends. The first factor is the number of stars in the galaxy. Although our estimate is uncertain by at least a factor of three, we shall adopt 400 thousand million for this number. Not all stars, however, are likely to have planetary systems. The stars that existed early in the history of the galaxy are believed to have formed from material with very low abundances of elements other than hydrogen and helium, and it is doubtful that those stars could have rocky planets capable of supporting carbon-based chemistry. This eliminates about half of the stars. Half of those remaining must be eliminated because about 50 percent of the stars, at least in our part of the galaxy, are in *binary-star systems*—pairs of stars revolving around each other at distances typical of the distances of planets from our sun. Planets could not have stable orbits in such systems. Thus we estimate that there are about 100 billion planetary systems in the galaxy.

Only some of those planetary systems, however, can have conditions suitable for the potential development of civilizations. Those whose stars are much more massive than the sun expend energy at a high rate and would exhaust their nuclear fuel and burn out in less than a billion years. We have no evidence for any life on the earth in its first billion years, and civilization took four and a half times as long to arise. Thus more massive stars than the sun have lifetimes too short to be likely to allow the rise of intelligent species.

Most stars, on the other hand, are much less massive than the sun and expend their energy very slowly. But they also shine feebly. For a planet revolving about such a star to receive enough heat from it to liquify water and allow the necessary chemical reactions it would have to have a small orbit and be much closer to its star than Mercury is to our sun. The tidal forces of the star would force such a planet to turn the same side always toward its sun, keeping one side very hot and the other very cold. All of its atmosphere would then flow around to the cold side, and promptly freeze.

In other words we expect life-supporting planets to be revolving about stars that are not too different from our own sun. Even allowing considerable leeway, at most a tenth of the stars are suitable. If each one with a planetary system had one planet at the right distance from it to be properly warmed, there would be about ten billion planets in the galaxy suitable for life. We shall tentatively adopt this number.

The next factor is the fraction of those planets on which life *can* develop on which, in fact, it *has* developed. Many biologists are of the opinion that given prebiological organic chemical reactions (forming, for example, amino acids), which seem almost inevitable, biological organisms will certainly follow. Obviously this assumption has never been verified. Should the Viking mission find organisms on Mars, either living or dead, it would strongly suggest that the fraction of suitable planets with life is, indeed, unity. For then we would know that life had arisen independently on two planets out of two suitable ones (and one—Mars—scarcely suitable at that). Lacking such information at present, let us suppose that the fraction lies in the range 0.1 to 1.0.

Living organisms, however, do not necessarily mean intelligent organisms with manipulative ability. Many biologists argue that normal biological evolution—mutations and natural selection—will inevitably lead to that stage. But on earth, only man has made it, and we required 4.6 billion years. What if we were unusually fast at it, and a more typical time required were four times as long, or even ten times? Then intelligent species would be found only rarely on planets with life. For the sake of illustration, however, let us assume that intelligent species develop on between 10 percent and 100 percent of the planets with life. Still, there is no assurance that an intelligent manipulative organism is interested in knowing about or communicating with similar creatures on other planets. Perhaps that human characteristic of curiosity is rare. Nevertheless, let us see where our estimates lead us if we assume that all intelligent organisms *are* communicative.

If all of the above estimates were correct, the number of planets in our galaxy on which intelligent, communicative beings of some description have evolved at some time or other would lie in the range 100 million (10^8) to ten billion (10^{10}). Now we come to the question that virtually all thinkers who dabble in this numbers game agree is the most uncertain: the longevity of a typical civilization.

Suppose a typical civilization remains in the communicative phase for a million (10^6) years. This time is about one ten-thousandth (10^{-4}) of the age of our galaxy—about 10^{10} years. Thus the above estimates for the number of civilizations that arise sometime or other must be reduced by a factor of ten thousand to find the number that are extant at this particular moment in galactic history. We shall return shortly to the reasons why civilizations would not be expected to last indefinitely. In fact, we shall see that our own civilization may be able to boast of a communicative technology for only one hundred years or so. One hundred years is only about 10^{-8} of the age of the galaxy. Suppose we combine the estimates of one hundred years and a million years for technological longevity, respectively, with our estimated range of 10^8 to 10^{10} planets that develop civilizations at some time. Then we obtain pessimistic and optimistic estimates for the numbers of civilizations currently available to communicate with. To be sure, the time for communication with a civilization remote in the galaxy may be long compared to the lifetime of that culture, but that fact does not affect the discussion to follow, provided that the total number of civilizations stays about the same even though individual ones may come and go.

We find, finally, that the number of planets in the galaxy with technological societies may lie in the range one (the earth) to a million. We could make the number even greater, but we would have to stretch the credibility of our estimates of the various probabilities more than many of us would be comfortable in doing. In any case, we should not take these numbers too seriously, for they depend on a string of assumptions concerning factors about which we have little or no knowledge.

The above discussion pertains only to our own galaxy. There are at least a billion galaxies in the observable universe, but even the relatively nearby galaxies are millions of light-years away. If we have any hope of learning about intelligent life elsewhere, it is probably limited to life in our own stellar system.

INTERSTELLAR TRAVEL

If there are one hundred other civilizations in our galaxy, the nearest one is probably more than five thousand light-years away. There would have to be ten thousand civilizations in order for there to be much chance of one being within one thousand light-years. Even if there were a million other advanced societies, we would not expect one of them to be as close as two hundred light-years.

Travel over such distances is not necessarily impossible. In fact, the United States has launched at least two spacecraft—Pioneers 10 and 11—that eventually will traverse interstellar distances. On the extreme long shot that either should ever be recovered by another civilization, each carries a plaque bearing drawings and symbols intended to supply information about us. (Actually, the purpose of the plaques was mainly to call *our* attention to the possibility of *other* civilizations. The spacecraft themselves, of course, would reveal far more about their makers.) The Pioneers, however, will require a very long time, hundreds of thousands or millions of years, to reach distances of hundreds of light-years.

Even at a speed of a tenth that of light it would take thousands of years for astronauts to reach the nearest civilization expected under the most optimistic estimates—and then only if they knew where it was. The energy required would be absolutely immense, and the spaceship would have to contain life-support systems capable of maintaining many generations of the colony of astronauts willing to subject themselves and their descendants far into the future to a space journey with an unknown destination and conclusion. We cannot rule out such a possibility, but it does not seem likely to come about in the foreseeable future.

On the other hand, it is well known that special relativity predicts time to pass more slowly than it does on earth in a system moving, with respect to us, at speeds near that of light. Thus, in principle, astronauts in a rapidly moving spaceship would age more slowly than we do on earth and could have time during their lives to reach a distant star.

Astronauts have not traveled fast enough to verify the slower aging of humans at high speeds. Nevertheless, this time dilatation predicted by relativity has been well tested experimentally. For example, cosmic rays, constantly striking the earth from space, break up some of the molecules of the upper atmosphere into multitudes of subatomic particles, among them mu-mesons or *muons*. But muons spontaneously disintegrate after an average time of about two microseconds. Even at the speed of light, an object can travel less than half a mile in this time. Yet, muons formed six miles up in the atmosphere are observed at the ground in large numbers. Because of the high energies of the primary cosmic rays (atomic nuclei) that formed them, many muons are moving at very nearly the speed of light. In *their own* time scales, they *do* break up in two microseconds, but relative to us their time is passing very slowly so they appear to survive much longer.

Table 1 shows how much time slows down, relative to earth time, aboard a hypothetical space vehicle moving at various speeds, expressed in terms of that of light. Also given are the numbers of years of crew time required to travel to a star one hundred light-years away and return to earth (but not counting acceleration and deceleration).

Table 1

Effect of Time Dilatation on Space Travel

Speed of spaceship in terms of speed of light (c)	Factor by which crew's time slows down compared to earth time	Number of years of crew time needed for round-trip journey to star one hundred light-years distant
0.9999c	70.712	2.8
0.98	5.025	40.6
0.95	3.203	65.7
0.90	2.294	96.9
0.75	1.512	176.4
0.50	1.155	346.4
0.10	1.005	1990.0

We see from the table that if astronauts could travel at 98 percent the speed of light they could make the round trip to a star one hundred light-years away and age only about forty years, although the family and friends they left behind would have long since been dead when they returned to earth. Unfortunately, the energy required to obtain such speeds is extraordinarily great. Suppose the entire life-support system required for the forty years of survival of the astronauts could be enclosed in a spaceship weighing only ten tons (three to five automobiles). We add another ten tons for engines, propulsion systems, and fuel. To give the vehicle a speed of 0.98c, the total energy required, no matter how rapidly or slowly it is expended, is about 4×10^{29} ergs—roughly the amount of energy expended by humanity worldwide, at the present rate, for two hundred years. To obtain such energy would probably require the complete annihilation of matter, which we do not at present know how to accomplish. S. von Hoerner has pointed out that if the crew wanted to reach its speed of 0.98c at an acceleration equal to the earth's gravity, which would take 2.3 years, it would need the equivalent of forty million annihilation plants of fifteen million watts each, producing energy to be transmitted (with perfect efficiency) by 6×10^9 transmitting stations of 10^5 watts each—and all of this apparatus must be contained within a mass of ten tons!

Interstellar travel is theoretically possible, but we see that the cost, the technology, and the psychological and sociological problems are formidable. We cannot rule out that man will someday travel to the other stars, or even colonize planets revolving about them, but it does not appear inevitable, and certainly not imminent.

HAVE WE BEEN VISITED?

Suppose other civilizations exist that have reached a technology that permits interstellar travel. Perhaps their individuals have lifetimes of many millions of years, or perhaps they have learned to control aging or to achieve suspended animation, as by freezing. Could such alien astronauts ever have visited earth?

While it would be difficult to prove positively that the answer to such a question is no, we must certainly dismiss the alleged evidence for such visits that fills much of the popular literature. Claims by certain authors that alien instruction was needed to build the pyramids of Egypt and other ancient structures are shown by archeologists to be completely unfounded, and, indeed, are based on half-truths and fabrications.

In fact, astronomers and astrobiologists would not expect us to have been visited. For one thing we cannot imagine how an alien civilization could know of our existence here. We have, to be sure, been inadvertently sending radio waves into space for the past several decades with ever increasing intensity; near the outer boundary of a sphere of electromagnetic radiation, growing with the speed of light, are waves coded with the humor of Jack Benny and Amos 'n Andy. But that sphere of radiation is less than fifty light-years in radius, and it is exceedingly unlikely that it has yet reached the nearest civilization, let alone with enough time to spare for them to dispatch spaceships back here to look us over.

Thus, if we have been visited or *are* being visited, the visitors are probably interstellar explorers looking at random for possible planets for habitation. Suppose we take our optimistic estimate of one million civilizations in the galaxy. If they were spread about evenly, there would be about thirty within one thousand light-years of earth. If we are typical, most of them would not yet have the capability of interstellar travel; but for the sake of argument, let us suppose that each does, in fact, launch one interstellar expedition per year to a star that could have a habitable planet. Now the number of stars within one thousand light-years is about 10^7, and according to our previous reasoning, about 3×10^5 of them would be candidates for habitable planets. Thus there is only about one chance in ten thousand of our being selected for a visit during any one year. Even according to our optimistic assumptions, we would not expect to have been visited even once during the time since man invented writing.

Many people believe that the so-called unidentified flying objects (UFOs) are spacecraft controlled by alien astronauts. There are at least several hundred reported sightings of UFOs per year, and some UFO-logists claim there are thousands. Probably some 90 percent are explicable in terms of known phenomena, but some lecturers on the subject insist that only about a tenth of the people seeing UFOs report them. If the extraterrestrial hypothesis for UFOs were correct, even counting return visits and multiple sightings of the same vehicles, we would have to assume that the earth was being visited by somewhere between twenty and several thousand interstellar spaceships per year. But even to be visited *once* each year would require—and *under extreme optimal assumptions*— that every galactic civilization launch ten thousand interstellar expeditions each year! Perhaps only very rare UFOs are extraterrestrial, and the others have different explanations; but if all save a very few can be explained in other ways, why not all? It is such reasoning that leads virtually all professional astronomers (perhaps *all*) to consider the extraterrestrial hypothesis the *least likely* explanation for the UFO phenomena—although not necessarily an absolutely impossible one.

Some UFO observers report actual contact with the occupants of strange spacecraft. According to J. Allen Hynek, director of the Center for UFO Studies in Northfield, Illinois, all such alleged beings have been humanoid, which either means that all intelligent creatures in our part of the galaxy are remarkably like us, or reflects a limitation to the imaginations of the reporters. One of the most famous accounts of contact with alien beings is that of Barney and Betty Hill, who reportedly were taken inside a spaceship and examined by humanoid creatures. The incident was recalled by the Hills only after hypnotic investigation, and the reality of it rests in large part on the authenticity of a star chart shown to Betty Hill by the aliens and later sketched by her from memory under a posthypnotic suggestion. Marjorie Fish, an Ohio schoolteacher, showed that the Hill sketch resembles the arrangement of fourteen nearby solar-type stars as they would be seen from a planet revolving about the star Zeta Reticulum, but astronomers Carl Sagan and Steven Soter have pointed out that the "agreement" is superficial at best, and is not statistically significant because of the freedoms of viewing angle and of selection of these particular stars from the Gliese catalog of nearby stars.

In short, it is the opinion of the overwhelming majority of physical scientists that the case has never been made for even a single UFO being of extraterrestrial origin; hard, unequivocal evidence is simply lacking. Whereas we cannot positively rule out that we have been visited by other civilizations, we have no firm reason to believe that we *have* been—not the kind of sober convincing evidence we would need to be able to assert proof of what would be one of the most exciting discoveries in the history of science: life beyond earth.

In fact, Michael H. Hart, of the National Center for Atmospheric Research at Boulder, Colorado, has started with the premise that we have *not* been visited by extraterrestrial beings to argue that we are the *only* technology in the galaxy. He argues that if there *are* other civilizations, at least some will have mastered interstellar travel and that at least some of these will have set out to colonize other habitable planets in the galaxy. Even allowing for considerable time between journeys from established colonies to new planets, such technological cultures would be expected to migrate through the entire galaxy in a time scale short compared to its age. We haven't run across such cultures; therefore they don't exist; therefore we are alone! Whether Hart's analysis is serious or tongue in cheek, it does present an effective devil's-advocate position to illustrate how easily we can turn arguments concerning the characteristics of other hypothetical civilizations in various directions to achieve conclusions we want to believe, and thus how fragile our estimates of the number of other civilizations really are. All the more reason to find ways of positively establishing the existence of life elsewhere. Is there a realistic way that this might be done?

EXTRATERRESTRIAL COMMUNICATION

It is just possible that remote civilizations might communicate with each other by means of signals coded on electromagnetic radiation. Astronomers vary widely in their opinion of the likelihood of this prospect. Most are cautiously skeptical, but do acknowledge the possibility, and all agree that it would be hard to exaggerate the importance of actual success in achieving such communication.

Even today, with our present technology, we have the ability to send radio messages across the galaxy with enough strength that equipment like that which we have in operation today would detect them. Two-way communication, however, is a long-term matter. At best we would not expect a signal we transmitted to be received in less than three hundred years, which means we would have to wait at least six hundred years for an answer. Still, we might eventually want to try transmitting messages to other stars; in fact our television programs are on their way to our neighboring stars anyway. To demonstrate the feasibility of transmitting radio signals over great distances, in 1974 the giant one-thousand-foot radio telescope at Arecibo, Puerto Rico, was used to transmit a binary-coded picture via radio waves toward the globular star cluster in Hercules, M13, some twenty-four thousand light-years away. It is unlikely that any of the hundred thousand or more stars in that cluster has an inhabited planet, but if a civilization there did have a similar telescope aimed toward earth it would be able (twenty-four thousand years from now) to detect our message.

Our only hope of knowing about other civilizations at present, however, is to *receive* messages from them. If such societies exist, hopefully they have reasoned along similar lines, and are broadcasting, as well as attempting to receive, messages. It is easy to devise codes that can be recognized as intelligently conceived; for example, binary-coded prime numbers or fundamental mathematical ratios such as pi (π) would be unmistakable. We would, however, have to guess what part of the electromagnetic spectrum would be used as carrier waves for such signals.

Such a guess is not difficult to make. The sun (and other solar-type stars) is extremely bright in visible light and in the infrared, compared to any radiation that we could transmit in those spectral regions. Thus visible and infrared signals from a planet probably would be swamped by the radiation from its parent star. Also, infrared and visible radiation, as well as ultraviolet radiation, is readily absorbed by the sparse distribution of dust in interstellar space, and would not penetrate more than a few thousand light-years through the galaxy in any case. Successful communication, therefore, will very probably be accomplished only at radio frequencies. At long radio wavelengths the galaxy itself is "bright" from synchrotron radiation produced by electrons in the interstellar gas spiralling in the magnetic fields of the galaxy. At short microwave wavelengths the earth's atmosphere is opaque. So at least our "listening" must be confined to radio waves with frequencies in the range from about 1 to 10 gigahertz (1×10^9 to 10×10^9 cycles per second).

Within this range of radio wavelengths we can concentrate a large amount of energy within a small bandwidth into a narrow beam that will carry thousands of light-years, and, at those wavelengths, will be many millions of times as intense as the solar radio radiation. Perhaps other technologies would reason along the same lines. Moreover, it happens that this radio "window" contains one of the most important spectral lines (at radio wavelengths) emitted in the galaxy—the radiation of wavelengths 21 centimeters (1420 megahertz) emitted by the neutral hydrogen atoms of interstellar space when their electrons spontaneously flip over so that they spin in opposite directions from the protons that comprise their nuclei. Observations of this radio spectral line by our own radio astronomers has revealed a great deal about the distribution of hydrogen in our galaxy, and hence much about galactic structure. Perhaps radio astronomers of other cultures, similarly aware of the astronomical importance of the 21-centimeter radiation, would select it as a natural frequency at which to try communication, for we (and others) would be more likely to pick it up by accident. In any event, if interstellar radio communication is going on, it has a good chance to be at frequencies of from 1 to 10 gigahertz—a spectral range not too great to monitor.

There have been at least eight more-or-less-organized programs to check possible extraterrestrial sources for coded radio signals. The earliest was Project Ozma, coordinated by Dr. Frank Drake at the National Radio Astronomy Observatory at Greenbank, West Virginia. Two stars were observed at a wavelength of 21 centimeters—Epsilon Eridani and Tau Ceti —both solar-type stars within twelve light-years. Three Soviet surveys monitored first twelve nearby sunlike stars and later the entire sky for pulsed signals at several wavelengths. Four other United States programs have been or are watching up to six hundred nearby solar-type stars, and even several external galaxies, for evidence of coded radio waves.

To date no signals from other civilizations have been detected. But then, no one really expected to detect anything from a sampling of, at best, a few hundred stars. With the most optimistic estimates we should have to survey ten thousand solar-type stars to find even one with a civilization; to encounter one beaming radio signals directly to us we might expect to have to monitor millions of stars. The surveys so far have been at most only "pilot" studies. To have a chance to detect other possible civilizations, surely we must survey all appropriate stars within at least one thousand light-years. Is such a search feasible?

One study that attempted to answer this question was made by a group headed by Bernard M. Oliver of Hewlett-Packard Company in 1971. The study, sponsored by Ames Research Center of the National Aeronautics and Space Administration, resulted in the design of a system called "Project Cyclops." The system consists of a vast array of radio-telescope antennas, each a hundred meters in diameter, spread over a region five kilometers across. The cost of the array was estimated at that time to be about ten billion dollars. Subsequently, Dr. Oliver has told me that he thinks that with the improved technology becoming available the system can be built at a much lower cost.

There are, at present, no plans to build Project Cyclops. If it were constructed, however, it would be able to monitor every solar-type star within one thousand light-years for intelligent signals within a period of about thirty years. Highly collimated beams of energy could be detected from much farther away. Cyclops could even "eavesdrop" on the equivalent of our own commercial television broadcasts originating from as far as a few hundred light-years away.

Two-way communication with other civilizations is, as was already explained, probably unfeasible. We can imagine civilizations thousands of light-years away that have ceased to exist by the time signals from them are received. Yet, there could be an entire network of technological societies scattered about the galaxy relaying information from one to another without any one pair ever engaging in a question-answer dialogue. The idea has been called "The Galactic Club." It is an imaginative and inspiring concept. This writer doubts that such a galactic club exists, but it would be one of the noblest endeavors of man to participate in it if it does. Thus many of us hope that Cyclops, or something like it, will be built. Even if it never detects another civilization, it will be a marvelous tool for radio astronomy and for our exploration of the universe.

LONGEVITY OF CIVILIZATION ON EARTH

The existence of technological civilizations elsewhere in the universe depends critically on whether such systems, once developed, can survive more than a few decades. In our own history wars, famine, and disease were at best perturbations on man's ascent in civilization. Not only were his means of self-destruction limited, but the earth was mostly undeveloped. Now man has the means to wipe out all living creatures on earth many times over in a few hours; also, the earth is full.

Perhaps the earth does not *seem* to be full. After all, there are still vast regions of the American Southwest that are unpopulated; and Alaska, Siberia, and Antarctica—to say nothing of the oceans—have thousands of empty square miles. Nevertheless, the earth *is* full, or nearly full, in its ability to provide food and resources for man's ever increasing appetite.

Consider New York City. With all of the problems of that great metropolis, surely no one would argue that it could support many more people. Yet there is less than one person for every one hundred square meters of land area in New York City. Imagine New York one hundred times as crowded as it is now—one person per square meter—and imagine a similar population density throughout the entire land area of the earth, including Siberia and Antarctica. At the present rate of its population growth the earth will reach that state of crowding in only 550 years.

The earth's population is doubling every thirty-five years—an annual rate of 2 percent. In some Latin American countries—Mexico, for example—the population doubles in only twenty years. Mexico must double its roads, hospitals, schools, homes, factories—everything it has— in just twenty years to maintain the same level of poverty most of its people enjoy today. World population growth is (almost) an example of geometric progression. For another example of a geometric progression, consider a standard sheet of typing paper, 0.1 millimeter thick. Suppose you cut the paper in two and lay one of the pieces on top of the other. Make a second cut, and stack the four sheets together; then you make a third cut, and pile up the eight pieces. Continue, if you can imagine it, until one hundred cuts have been made. The stack of paper would then be ten billion light-years thick! This is a distance comparable to the most remote quasar ever observed!

But the earth's population is not increasing *exactly* at a geometrical rate. Actually it is a bit faster; the rate of population increase is itself proportional to the population. If this rate could continue, the population would become infinite in a finite time. The present growth formula leads to an infinite population in 2026—in fact, on Friday, November 13, a day dubbed "doomsday."

Of course the earth's population will not become infinite, for mass must be conserved. The trouble is, it increases so rapidly that disaster from a less than infinite population is upon us before we realize what is happening. Our present rate of energy use is increasing at about 7 percent per year, further compounding our problems. Our stockpiles of weapons are increasing exponentially as well; even now we have nuclear bombs stockpiled in an amount equivalent to ten tons of TNT per person—enough to raise a ten-thousand-ton building five hundred meters into the sky above each man, woman, and child in the world and then drop it on him. The queues at gasoline pumps many of us "suffered" through in 1974 were nothing compared to the queues and even riots we can expect in supermarkets when our food supplies run short.

The ideas of Thomas Malthus are not a theory—they are a cold mathematical fact in a society with an increasing population on a finite planet. There is only so much matter on the earth. Regardless of how you transform it, it can support only so many people. Nor will colonizing space help. Even if we wished to send our children to Mars and the moon to live (without fields to play in, rivers to fish in, and lakes to swim in), and could make habitable dwellings on those worlds as fast as needed, we would gain only one generation—thirty-five years—for the combined area of those worlds is only about that of the land area of the earth. If we could colonize every possible planet around all possible stars at the density of people the earth has now, all such planets within 150 light-years would be filled within five hundred years, and to find room for our ever-increasing population we would have to transport people to new planets at speeds greater than that of light, which is impossible.

In short, we cannot increase our population without limit. It *will* limit itself as space and supplies run out, but what will life be like then? Can we imagine that all of the ills that go with overpopulation—crime, poverty, famine, disease—will not make life here a living hell before humans snuff themselves from existence? Our only rational solution would seem to be to limit our own numbers—which hardly seems to be in the offing, worldwide. But even if we could do so there would remain problems, such as genetic deterioration (medicine increases the life span but encourages survival of persons with genetically unfavorable mutations), and the ultimate crisis of the boredom and stagnation of a stable society trying to endure without substantial innovation on a completely filled planet for hundreds of centuries.

Such are the arguments that our society may be doomed to a short longevity—perhaps only one hundred years or so of technology. Have we misread the clues? Are there alternatives? Let us hope there are. The positive evidence, however, is not at hand. What more important reason, therefore, to discover even *one* other civilization that has endured for tens of thousands of years; at least if those fellows made it, perhaps there is hope for us!

If for no other reason, this would give us good grounds to search for, and to hope to find, other societies that have surpassed us, not only technologically but sociologically as well. The search for life in the universe will not be easy. It may never succeed. In particular, the matter is of too great import to give credence to the unfounded claims of pseudo-scientists and charlatans. On the other hand, the true discovery of life in the universe would not be just a milestone for man; it might well be his salvation.

How often have we been led to believe that first encounter with ETIs will inevitably mean war, or worldwide panic, or wholesale disaster, or The End—or something of the sort? H. G. Wells' *War of the Worlds*, Orson Welles' famed radio performance of that classic in 1938, plus endless television sci-fi plots (e.g., the British series "UFO") all carry the heavy-handed message that the discovery of intelligent alien life must be kept a military secret and hidden from the public at all costs.

Kendrick Frazier, a distinguished science-news reporter, here warns that such an attitude is dangerous in the extreme. He contends that honest dissemination of information about such contact, handled with foresight and maturity, could turn it into one of "the most momentous events in the history of humankind."

Frazier's major interests are "earth, oceans, atmosphere, astronomy, space exploration, philosophical and social issues of science" ("major interests" only, note), and he has reported on almost every significant event in these fields in recent years. He covered American scientific-research programs in the Antarctic (1973) and did the photographic layout for *The World Book Science Annual* (1975).

Ken Frazier is the editor of *Science News*, the most widely read science-news source in America.

"It is not too early, I think, for the NSF to begin considering the funding of a study of the human consequences of contact with intelligent extraterrestrial life. The study might be premature by decades or centuries. But then—it might not be."

Kendrick Frazier

First Contact: The News Event and the Human Response

On a drive back from a lonely area in western New Mexico in the summer of 1974, I began wondering about the subtleties of what had drawn me there. I had gone to the Plains of San Augustin, formerly a broad Pleistocene lake bed, dotted only with occasional ranchhouses and windmills, to see for myself the site of what in a few years will be the most advanced center for radio astronomy on earth. By 1981 no fewer than twenty-seven radio-telescope dishes, distributed over the terrain in a Y-shaped pattern, each arm of the Y thirteen miles long, will be aimed in unison at the heavens and linked electronically into one grand radio receiver far more sensitive than any yet built. Someday, I'm sure, this superstar of radio-astronomy observatories will be given a proper name, but for now it is inelegantly called merely the Very Large Array.

In a few years this series of radio dishes will cut a striking figure on the regional horizon, but now there is nothing there but a trailer, some earthmoving equipment, and a few paths scraped through the grama grass and sandy soil by a road grader. Construction of the VLA had not yet started. And yet I was entranced. I was struck with a strange exhilaration, an almost spiritual feeling of being in the presence of immense mystery and awesome, incomprehensible potential.

It was, I eventually realized, the same kind of feeling I had had when I visited large radio-astronomy observatories that were already operating at Green Bank in West Virginia and at Arecibo in Puerto Rico and even at smaller ones in Maryland, Arizona, and Colorado. What was it that had so attracted me to all these places? I knew it wasn't an interest in the antennas themselves, although they do have a certain grandeur, or in the electronic equipment, sophisticated as it may be. I care little and know less about such things. Was it merely my love of astronomy, my appreciation of any new instrument that helps us better explore the marvelous intricacies of the universe? Perhaps.

The more I thought about it, however, the more I became convinced that there was something more basic involved, something with deep philosophical meaning. When it came, the answer seemed obvious. Here at one of these radio observatories or at one of their successors may occur the most momentous event in the history of humankind—the arrival of the first communication from an extraterrestrial civilization. Here may begin the next New World. Here may come the patterned electromagnetic ripples that will reverberate throughout human society, shaking our social and political institutions to the core, revolutionizing the way we view ourselves, and transforming our world into a new consciousness, for better or for worse. Here is our direct link to the intelligent cosmos, awaiting only some galactic S. F. B. Morse to sit down at his end of the connection and tap out his civilization's version of "What hath God wrought!" What indeed!

"The assertion that 'God created man in his own image' is ticking like a time bomb in the foundations of Christianity."[1] So wrote Arthur C. Clarke in assessing as slight the chances that very many of our current religions would survive the first message that brings proof of a greater cosmic intelligence. I have taken those words of Clarke as the gospel on the subject ever since I first read them more than ten years ago. Our religions are not the only of our institutions built upon an unjustifiably self-centered sense of our own importance, so it seems reasonable to anticipate similarly shattering effects throughout broad parts of society as a result of the confirmation of extraterrestrial intelligence.

Yet before taking for granted those assumptions and going on to explore their implications, let's not too hastily discard the notion that society can accommodate even that kind of realization. Take religion, for instance. The church that once persecuted Galileo and Bruno for then-unorthodox views about the universe has, in the form of an address by Pope Pius XII to the International Astronautical Congress in Rome, explicitly endorsed the idea of manned interstellar spaceflight as a laudable goal for man—"so that his spirit might gain an ever more profound knowledge of the infinite greatness of the Creator."[2] It seems only a small additional step to rationalize that the existence of intelligent civilizations elsewhere in the universe is only a further manifestation of "the infinite greatness of the Creator." What I am trying to emphasize is that institutions, especially those with a good record for endurance, have a certain amount of stretch built into them; thus it may be an underestimation of their resiliency to expect them automatically to crumble at the first hello from elsewhere in the galaxy.

Many theologians and humanists who exalt the special status and uniqueness of humankind on earth see no contradiction in holding to that view while also acknowledging the virtual certainty of there being more intelligent beings elsewhere. All that is required is a shift in emphasis from, as Harvard Divinity School Dean Krister Stendahl puts it, "absolute uniqueness" to "our more relative uniqueness." The situation may have been different ten or twenty years ago, but I have a feeling that today one would find considerable support among the enlightened segments of the theological community for Stendahl's attitude of enthusiastic welcome of communication with life beyond earth: "It seems always great, to me, when God's world gets a little bigger and when I get a somewhat more true view of my place and my smallness in that universe. The growing awareness of cosmic cohabitation is enormously important for me, and it fits well into a growing knowledge of God's world."[3]

Another reason that confirmation of extraterrestrial intelligence might not necessarily have as disruptive an effect as we might expect is that considerable cultural conditioning has already taken place. The high likelihood of there being other worlds similar to ours, based on the numbers of stars and galaxies astronomers can see, has, after all, been an accepted fact for many decades. "Since at least the 19th century it has been considered the height of philosophical naiveness in scientifically sophisticated circles to believe that the earth is something extra special in the universe," reminds molecular biologist Gunther S. Stent.[4] More than two decades have already elapsed since Stanley Miller's classic laboratory demonstration of the ease with which the chemical precursors of life can be produced from the primordial soup. The more recent laboratory experiments in chemical evolution, observations of organic chemistry between stars, and confirmation of amino acids in meteorites have further increased the odds for the existence of extraterrestrial life. But it could be argued that these are only the icing on an already baked cake of evidence pointing to the inevitability of intelligent life elsewhere in the universe. Thus most educated people have been exposed in some degree to the notions we are talking about. Opinion leaders in the social sciences, arts, humanities, and even politics have had or, by the time it becomes crucial, will have had, time to get used to the idea that we are not alone. (In fact, an interesting side issue is whether the discovery of extraterrestrial intelligence will be anticipated to such a degree for so long that a failure to find evidence for it after a search of many decades might produce a philosophical backlash that would undo all the beneficial cultural conditioning that had gone on before. The proposition that we really are alone in the universe may be far more frightening than its contrary.)

So some of our institutions may be able to stretch or bend enough to adjust to the realities of the existence of extraterrestrial life, and our cultural and public-opinion leaders may be able to help the general population get used to the idea.

But despite all the institutional flexibility and psychological preconditioning in the world, it is hard to imagine that we could ever be fully prepared for the shock, whenever it comes, of the first confirmed message from other beings. This does not mean chaos is inevitable; it just means that there is no way even to predict, let alone prepare for, all the implications of the event. Even years after it has happened we will probably be unable to anticipate the long-term significance in any accurate way.

The first European accounts of Columbus's discovery of the New World were devoted mostly to gold, the naked natives, and the opportunity to convert them. Columbus himself died ignorant of what he had really discovered, and Samuel Eliot Morison tells us that "of the real significance of the discovery for Europe's future there was not one hint in contemporary comment."[5] Of course we might pride ourselves on being far more knowledgeable and sophisticated than Columbus's contemporaries, yet the earth's first contact with an extraterrestrial civilization seems an event of far greater magnitude and with a much greater likelihood of unknowable significances than late fifteenth-century Europe's first contact with a new world across the Atlantic.

I have been discussing the discovery of the existence of extraterrestrial intelligence as an event, as though it is destined to come in a single, dramatic moment. One day the earth is an isolated, unaware child in the wilderness; the next it is part of a vast cosmic brotherhood. That may prove to be the case. Many discoveries in astronomy have happened that way—with sudden, clear-cut decisiveness. But it will not necessarily be that way, and all references to "event" might best be considered a convenient metaphor for what may be a series of incremental steps over an extended period of time.

I have long been fascinated with exactly how the first signal will be detected, identified, and confirmed. Will that process be rapid—a matter of days or, at most, weeks? Or is it likely that years or decades might pass between the time some anomalistic radio signal is received and the time it is eventually identified and confirmed to the satisfaction of the world scientific community to be the manifestation of extraterrestrial intelligence?

The longer time interval is perhaps nearly as likely as the shorter, and just as interesting to contemplate. The message may not be recognized as a message. Or it may be recognized as a possible message but not be susceptible to confirmation. Maybe it will prove to be so technically complex or based on such vast cultural differences that interpretation will have to await a new generation of computers or a newer and wiser generation of human minds. The situation is full of subtleties, and I have always thought it would make an interesting science-fiction story. (Having said that, I await word of the final total of how many such stories have already been written. I am a great admirer of science fiction, enough so to postulate that for any novel idea one might think of, there has already been written a science-fiction story using that idea as its theme.) Let us pose a few trivial examples of the possible implications of the extensive-delay-to-confirmation situation: scientific reputations would rise and fall; whole cultist philosophies based on various expectations about the aliens would have time to come and go; political parties based on opposing beliefs about the extraterrestrials could be formed, do their good or dirty work on the public mind and pocketbook, and fade away, leaving who knows what in their wake; systems of public and higher education could go through several cycles of restructured approaches based on shifting philosophies. All of this could happen between the first suspicions of positive evidence of intelligent life elsewhere and the confirmation of that evidence.

During this extended period society would be in a psychologically

ambiguous situation. How should society react to an event of unprece-
dented importance in human history, when it could not be certain that the
event had really happened? A new and uncertain variable would for a
period of time be having an influence on human attitudes and beliefs.
Such uncertainty would cause enormous problems, but at least it would
have the advantage of serving as a kind of society-wide trial balloon to see
how people, groups, and social organizations react and change as a result
of just the possibility that the momentous communication had been received. This would enable better preparation for the social effects of its
confirmation. The situation of possible-but-unconfirmed contact would
also allow people the luxury of a more gradual adjustment to the new
realities.

As a science newsman, I have always had a particular interest in the
way the first news of extraterrestrial intelligence might be reported to the
world. The accuracy and tone of the initial coverage will determine to a
large degree the way the people of the world will react during the first few
days and weeks. It could therefore go a long way toward shaping the
psychological and philosophical attitudes of the populace for years to
come. Factually inaccurate initial reports would give rise to mistaken
ideas, wild rumors, and confusion. A tone of hysteria in the news reports
would contribute to hysteria among the public. My hope is that from the
very beginning there would be accurate and restrained news reports,
complete with all the facts available at the moment and including the
comments, insights, and perspectives of all the scientists and other persons
informed and knowledgeable on the subject.

My personal nightmare in this regard is that the apparent discovery
will be made not in the United States or Great Britain, for example, but in
a country with no tradition of openness and candor in the release of news
information. The worst thing that can happen in a situation where a
major news event with enormous possibilities for disruption of people's
values and beliefs occurs is for there to be no reliable continuing source of
accurate and complete information. A brief and incomplete initial report
followed by a moratorium, intentional or not, on release of further information would set the stage for great public misunderstanding.

Even if there were reasonably complete release of initial information
from a country without a tradition of openness, say the Soviet Union, it
would still be difficult for scientists and the press in the Western world to
evaluate its reliability without a better tradition of quick and open communication than now exists. Already there have been a few claims coming

out of the Soviet Union of receipt of radio signals that were initially attributed to extraterrestrial intelligence. (They turned out not to be.) The difficulty that both the science press and scientists in the United States faced in getting further information from the Soviet Union in the days immediately following the initial news reports has not made me especially confident.

I would hope that full release of all scientific details would quickly be made to the world scientific community through international scientific organizations, so that scientists everywhere could have access as soon as possible to the technical data for their own evaluation. Their assessments could then be quickly relayed to the news media, to alleviate or shorten any gap between the initial, perhaps distorted, reports and more authoritative judgments based on the evaluations and reflections by scientists of diverse backgrounds and perspectives.

Notice that, despite my great concern about the effect of the discovery on the public mind and about the need for sober and responsible news reports, I do not support the idea that the information should be withheld from the public because of the notion that "the public is not mature or wise enough to cope with its implications," or some such rationalized nonsense. The public—that is, us, all of us—is not mature and wise enough to deal with any of a variety of problems, issues, crises, and concerns that face the world today, but we all have to try. And we lose no nobility for the effort. That is the tradition of the human race. Facts are facts and truths are truths, and hiding them or hiding from them is futile. History is replete with examples of the damage caused by trying to withhold important information from the public. Anyway, practicalities aside, the idea is morally and philosophically repugnant. We, or our descendants, all will have to grow up together amidst the realities of a world forever changed by the certain knowledge of a greater extraterrestrial intelligence, and when that time for the world arrives there will be no use trying to postpone the pains of adolescence. It will truly be, to borrow the title of one of Arthur C. Clarke's books, childhood's end.

Wherever the discovery of the first evidence of extraterrestrial intelligence is made, I do hope that the scientific leaders involved will be wise enough to prepare their public announcement with a certain amount of care and with careful attention to the need to supply accurate information and perspective. This would justify a delay of several days to make preparations. In the United States a joint news conference in Washington, with scientists and officers of the discovering observatory, the

National Science Foundation, the National Academy of Sciences, and the White House, would perhaps be the preferable method of announcing the discovery. The international scientific community would be brought into the matter as rapidly as possible.

It is interesting to note in this connection that a scientist who was for many years the foreign secretary of the National Academy of Sciences and who was in 1974 elected president of the International Council of Scientific Unions, Harrison Brown, is himself the coauthor of a novel based on the first contact with extraterrestrial intelligence. (The book, *The Cassiopeia Affair*,[6] deals with national and international political events following the receipt of a message from near the star Cassiopeia 3579 at an American radio-astronomy observatory in the 1990s. The news is soon brought to the attention of the president, who, like the observatory director, is an idealist, an internationalist, and a proponent of world peace. The president announces the news to the world. But when he starts capitalizing on it to help bring about his dream of world disarmament, his political enemies begin attacking the validity of the message, which was received on an evening when the observatory director was alone in the laboratory. This leads to congressional hearings and accusations of fraud. The scientist's credibility is undermined, he dies of a heart attack, and the discovery is discounted. As the book closes, the same precisely pulsed message from Cassiopeia 3579 is received by radio astronomers in China.)

It is a hope of mine that the discovery of extraterrestrial intelligent life will come during my professional lifetime and that I might play some role in the reporting of the event and the interpretation of its significance. But whether or not that dream comes to pass, one can only hope that those who communicate that momentous information recognize the importance of establishing from the very beginning a tradition of candor, factual accuracy, and responsible comment. It will be important to separate facts and responsible speculation from fantasy and irresponsible speculation, to make as clear as possible the distinction between relevant concerns and irrelevant ones, to hold the discussions from the beginning on a plane of respectability effectively out of the reach of the pseudoscience fans and cultists, who no doubt will be out in force to exploit people's natural latent fears and lack of sophistication in dealing with scientific concepts. (The UFO experience has not been encouraging in that regard.)

An event that offers some useful parallels is the launch of the first artificial earth satellite by the Soviets in 1957. The event took the American public and most United States government officials by surprise. It caused a mixture of admiration, awe, and fear. The impact that the Sputnik launch had on the American educational system, on expenditures for basic science, and on the mobilization of an American space program are well enough known to not need repeating. But those were longer-term effects. In the first days and weeks after the launch, some of the response was near hysteria. Fortunately the scientific directors of the Smithsonian Astrophysical Observatory, which was tracking Sputnik, were profoundly convinced that the public was entitled to know everything.[7] They established a policy that no question would go unanswered if an answer could be found. The observatory became the information source for the entire Western world on the strange new object in the sky. Reporters slept overnight at the observatory's offices at Kittredge Hall in Cambridge. By issuing the latest information and making its scientists available for comment the observatory did much to bring about news coverage that was accurate and reassuring and to minimize coverage that would arouse further fear and anxiety. Newspapers confronted with wild rumors began checking with the observatory before printing the reports, usually finding that they were unfounded or greatly exaggerated. All in all, recalled the observatory's director of publications years later while reviewing the thousands of newspaper clippings on Sputnik in the observatory's files, "one is again and again impressed with the accuracy and the thoroughness of the reportage."

I referred earlier to the difficulty of predicting the human response to and assessing the eventual social implications of the discovery of extraterrestrial life. While that is true, there are a handful of social scientists in the United States whose specialty is attempting to anticipate or to assess the human response to major natural disasters or to new technological capabilities. The human events surrounding hurricanes, tornadoes, floods, and earthquakes have in the past been the subject of such studies. More recently the possible consequences of weather modification and earthquake prediction have been studied. It is not too early, I think, for the National Science Foundation to begin considering the funding of a similar study of the consequences of contact with intelligent extraterrestrial life. The study would try to assess public attitudes toward the almost certain existence of extraterrestrial intelligence and the possible public response to the actual discovery of it. In light of what we now know about chemical evolution, the first half of the study would certainly be timely today. The second half might be premature by decades or centuries. But then—it might not be.

For the apparent irregular movement of the planets and their variable distances from the Earth . . . make it clear that the Earth is not the center of their circular movements. . . . The order of the spheres will follow in this way—beginning with the highest: the first and highest of all is the sphere of the fixed stars. . . . Saturn, the first of the wandering stars, follows; it completes its circuit in 30 years. . . . After it comes Jupiter. . . . Then Mars. . . . The place fourth in order is occupied by the annual revolution in which we said the Earth together with the orbital circle of the moon as an epicycle is comprehended. . . In the fifth place, Venus. . . . The six and final place is occupied by Mercury. . . . In the center of all rests the sun.
—Copernicus, *De revolutionibus orbium coelestium* (1543)

When on board H.M.S. *Beagle* as a naturalist, I was much struck with certain facts in the distribution of the organic beings inhabiting South America, and in the geological relations of the present to the past inhabitants of that continent. These facts . . . seemed to throw some light on the origin of species. . . .   —Charles Darwin, *The Origin of Species* (1859)

These two great intellectual discoveries have shaped our world. Copernicus showed that the planets, including our earth, revolve around the sun, thus rejecting our special status in the physical universe. Darwin showed that all life, including human beings, evolved from earlier creatures, thus rejecting our special status in the world of life on earth.

The realization that there is intelligent life elsewhere in the universe unites these two seminal ideas and carries them a giant, but probably inevitable, step further, to the level where the inanimate macrocosm—the world of galaxies, stars, and solar systems—meets the organic microcosm —the world of organisms, cells, and replicating genetic material. It is a situation I find fitting and altogether satisfying. Nature, after all, has no artificial boundaries, and the quest to understand its mysteries is gloriously advanced by conceptions that unify previously disparate parts.

Whether or not we humble beings, on our lonely planet, in our relatively isolated segment of the galaxy, ever manage to monitor the electromagnetic tracings of intelligent beings elsewhere, we can be reasonably certain that those beings are there. How each of us reacts to that realization probably is not much different from the way each of us reacts to any of a variety of other new inputs to our consciousness. To me and, I think, to many other people the idea is deeply fascinating and full of possibilities for a greatly enlivened future for humankind. To others it is a fearsome threat, full of psychological and other hazards for humans' sense of their own self-importance. To still others it is interesting but irrelevant; to them, the world is faced with such enormous problems now that there is not much point in even thinking about the more distant future.

This latter attitude seems particularly prevalent today, perhaps with good reason; so I wish to dwell on it a moment. Shortages of energy, food, and resources, a shaky world economic system, overpopulation, overpollution, and the continuing threat of nuclear devastation are hardly trivial concerns. How can one be contemplating the uncertain exigencies of extraterrestrial life when the very real problems facing many humans need so much attention? It is an honest question, one that is often asked of those in the field of science. The answer requires no criticism of either point of view. It is that we need to do both. To state it in personal terms: to me, the discovery of some previously unknown principle of nature, such as the inner structure of the proton or the shape of the universe or an ingenious evolutionary adaptation of a butterfly, is an event filled with wonder and beauty and majesty. I am not a scientist, but I can share with the scientist his sense of pleasure at learning something new about the marvelous workings of nature. So, too, a compassionate and caring friend is rewarded by any step that might aid a starving person in a drought- and famine-plagued Asian country. We need both kinds of endeavor to be a truly civilized nation. A society is ennobled by its ability to attend simultaneously to the solution of the practical concerns of the moment and to the creation of an enriched philosophical legacy for the future.

While some of us are devoting our creative energies and resources to solving today's problems, others of us can be looking ahead to tomorrow's possibilities. That is the way it should be. That is the strength of a vital society. That is the diversity of inner drive that makes the human species such a versatile and adaptive creature, despite its undoubted relative insignificance in the cosmic scheme of things.

Like mankind's first view of earth from lunar orbit, the realization of our brotherhood with other beings elsewhere will expand and improve our perspective on our place in nature. That it may prove unsettling we can hardly doubt. But rather than look upon that possibility with trepidation, we should look upon it with anticipation. The questioning of ideals, beliefs, and attitudes and the infusion of new perspectives is part of the creative process, part of the revitalization that brings growth and vigor to any culture.

When the certainty of extraterrestrial intelligence becomes impressed on the human consciousness, it will bring a new renaissance to human culture. Today, as we send unmanned probes to the planets of our solar system, some of our earth-oriented physical sciences—especially geology and the atmospheric sciences—are beginning to undergo a tremendous transformation and expansion as they reap the scientific advantages of direct knowledge about the surface and atmosphere of Mars, Venus, Mercury, and Jupiter. Comparative planetary geology—the term has an intriguingly adventurous ring to it!—is now a reality. Perhaps, just perhaps, the science of comparative planetary biology will be born soon after the first unmanned spacecraft, equipped to test the Martian soil for the presence of organisms, lands on Mars in the summer of 1976.

But the consequences of these infusions of new knowledge into the physical and biological sciences pale in comparison to the ways the disciplines of sociology, anthropology, psychology, and linguistics would be invigorated and transformed by the realization of the existence of extraterrestrial civilizations. These fields thrive on the comparison of social groups and cultures. Who can even imagine the fascinating perspectives that will be gained from the first communication with a culture from elsewhere in the universe? Will we find that what we now consider to be enormous differences between societies here on earth are insignificant in comparison with the differences between earth culture and the culture of a civilization on another planet? Will this make us reassess our views toward our fellow human beings? Will the thought patterns of other beings elsewhere have anything in common with ours? Will their language even be recognizable as language? What lessons will their governments and social systems have for us? Will they have solved any of the social problems that have afflicted human societies for thousands of years? Was there something in their makeup that prevented destructive conflict from the beginning? Have they evolved intellectually and culturally faster than we, or have they just been at it longer? One could go on asking questions. The interesting thing is that no matter how many questions we ask, many of the questions that finally come will be ones that we never thought of.

Similarly the effects on art, literature, poetry, and music will be enormous. Our understanding of the physical laws of the universe is certain to take a quantum jump. Our technology should reap benefits as well. But I deemphasize that because of my strong feeling that we need to realize that the physical and intellectual exploration of the universe, and the quest for contact with whoever is out there, is an intensely human undertaking, not merely a technological enterprise. It goes to the roots of our need to understand ourselves and to push back the borders of knowledge and expand the frontier beyond. It is a glorious endeavor of humanity.

Pushing back the frontiers of space to whatever new world waits beyond is not, in the most fundamental sense, the enterprise of the technical expert—despite the unfortunate flavor of technocracy that tended to emerge from the first big space-exploration effort to touch the human consciousness in a broad way, the moon-landing program. Nor can the real product be tangible. I cry out in agreement with Ray Bradbury when he recalls with dread "all the times . . . when I heard NASA apologists applying poultices to the tax wounds of the American public by describing how the Apollo missions have given old ladies new plastic cooking pans made from nose-cone chemistries and attached more proficient bumpers to mindless motorcars."[8] Obviously those justifications miss the whole point.

The challenge of the universe out there is not just for the scientists and engineers; it is for all of us. It is for our spirit, our soul, our psyche— for whatever it is that makes us what we are. It is for our poets, our philosophers, our dreamers, our artists, our humanists. It is the intellectual and philosophical adventure of our time. That would be true even if the universe remained to us forever inanimate. But someday we will hear that precisely pulsed message that verifies our membership in a vast and living community of civilizations. And we will never again be the same.

NOTES

1. Arthur C. Clarke, *Profiles of the Future: An Inquiry into the Limits of the Possible,* rev. ed. (New York: Harper & Row, 1973), p. 94.

2. Eugen Sanger, "Beyond the Solar System," in *The Coming of the Space Age,* ed. Arthur C. Clarke (Des Moines, Iowa: Meredith, 1967), p. 216.

3. Krister Stendahl, in *Life Beyond Earth and the Mind of Man,* ed. Richard Berendzen, National Aeronautics and Space Administration (Washington: U.S. Government Printing Office, 1973), p. 29.

4. Review of *Life: The Unfinished Experiment,* in *The New York Times Book Review* (Sept. 8, 1974), p. 27.

5. Samuel Eliot Morison, *The European Discovery of America: The Southern Voyages* (New York: Oxford University Press, 1974), p. 98.

6. Chloe Zerwick and Harrison Brown, *The Cassiopeia Affair* (Garden City, N.Y.: Doubleday, 1968).

7. See E. Nelson Hayes, "Tracking Sputnik I," in *The Coming of the Space Age,* p. 5.

8. Ray Bradbury, *Mars and the Mind of Man* (New York: Harper & Row, 1973), p. 141.

All of us wonder, sometime or often, what it might feel like to be someone else; and we science-fiction addicts *all* fantasize, in our wilder moments, what it might be like to be an ETI.

Well, no TI has come closer to becoming an ETI than Leonard Nimoy, who played the half-Vulcan, half-human Mr. Spock for three seasons in the television series "Star Trek." In this article for *ETI* he "reassembles" some aspects of that experience. So closely did he identify with the alien Spock that Nimoy recently found it beneficial to write a fascinating disclaimer (for others, yes; for himself, maybe—?), which he emphatically entitled *I Am Not Spock* (1975).

Leonard Nimoy is an actor, producer, director, recording artist, photographer, and author. On the stage he has starred in *Fiddler on the Roof, Visit to a Small Planet, Oliver, Sherlock Holmes,* and many other plays. His motion-picture credits include *The Balcony, Deathwatch, The Alpha Caper, Assault on the Wayne,* and *Catlow*. On television he has starred in two epochal series, "Star Trek" and "Mission: Impossible."

Mr. Nimoy is the author/photographer of *You & I* (1973) and *Will I Think of You* (1974).

"For three years, twelve hours a day, five days a week, approximately ten months of each year, I lived the life of an extraterrestrial. I began to study human behavior from an alien point of view. I was becoming alienated, and I didn't realize it."

Leonard Nimoy

Conversations with Mr. Spock

NIMOY: Spock, what is life?

SPOCK: A state of being.

NIMOY: Let me put the question another way: Why is there life?

SPOCK: Yours or mine?

NIMOY: Anyone's.

SPOCK: You've missed my point.

NIMOY: Which?

SPOCK: "Yours or mine." I was trying to suggest something. . . .

NIMOY: You've lost me.

SPOCK: How can that be? To have lost something is to be unaware where it is. You are still here.

NIMOY: It's just an expression. I meant I don't understand.

SPOCK: Your question is based on a personal conjecture. "Why is there life?" suggests that there is a single reason for all of life.

NIMOY: Is that wrong?

SPOCK: Is that necessary?

NIMOY: Many people are troubled by this question. They feel somehow that their lives would be enriched if they knew why they are here.

SPOCK: Are you religious?

NIMOY: I think so.

SPOCK: Then perhaps you are here because God wants you to be.

NIMOY: [*stunned*] Spock, I'm really surprised to hear you say that. Do you mean it?

SPOCK: Why does that matter? We are investigating possibilities. Does that one appeal to you?

NIMOY: Possibly. At times.

SPOCK: Do you believe in the concept of service to mankind?

NIMOY: I think so.

SPOCK: Then perhaps you are here to be of service.

NIMOY: Suppose neither of those possibilities satisfies a particular individual?

SPOCK: We've only suggested two. Aren't the possibilities endless? And why do you feel responsible for supplying the answer for everyone? Have you chosen that as your reason for existence?

NIMOY: Then each person has to find his own?

SPOCK: "Has to . . . " That would seem to suggest an emotional need, to which I find it difficult to relate.

NIMOY: Are you going to hide behind that mask of logic?

SPOCK: Are you angry with me?

For three years, twelve hours a day, five days a week, approximately ten months of each year, I lived the life of an extraterrestrial.

Many people have had some experience with role-playing. Some people have had more experience than I, and some have played a particular role longer than I. But given my intense commitment to identification with the role, and given the unique nature of the role of an "extraterrestrial," there may be some value in reassembling the experience.

Six years after having completed the role, I am still affected by the character of Spock, the Vulcan first officer and science officer of the Starship Enterprise. Of course, the role changed my career, or rather, gave me one. It made me wealthy by most standards and opened up vast opportunities. It also affected me deeply and personally, socially, psychologically, emotionally. To this day I sense Vulcan speech patterns, Vulcan social attitudes, and even Vulcan patterns of logic and emotional suppression in my behavior.

What started out as a welcome job to a hungry actor has become a constant and ongoing influence in my thinking and lifestyle. In 1965 Gene Roddenberry was producing a TV series for NBC titled "The Lieutenant." It starred Gary Lockwood and had to do with his adventures as a Marine Corps officer. I was hired for a guest starring role.

A year later, as a result of that and other TV performances, Roddenberry cast me in the role of Mr. Spock, son of Sarek, the Vulcan, and Amanda, the earth scientist, in the science-fiction adventure series "Star Trek." The Vulcans had been a violent and emotional people, which almost led to their destruction. They made a decision. Thenceforth emotion was to be foreign to the Vulcan nature. Logic would rule. Vulcans would be distinguished in appearance by their skin color, hairstyle, and pointed ears. They would be a race concerned with dignity and progress, incorporating the culture and ritual of the past with the best of what the future could offer. In Spock there would be a special mixture of tensions—the logic and emotional suppression of the Vulcan people through the father, Sarek, pitted against the emotional and humanistic traits inherited from the human mother, Amanda.

Thus we gave birth to an extraterrestrial.

The first labor pains took place in a makeup room at Desilu Studios in Hollywood. I sat down in front of a mirror, and Lee Greenway, a durable makeup man and old acquaintance, started to experiment with an admittedly crude application of the first pair of pointed ears. They were "built up" with layers of paper tissue and liquid latex and were never expected to be acceptable, but only to give an indication of what the effect might be. Well, it was as bizarre as the result of a child's playing with its mother's makeup. It was gruesome, ludicrous, and very depressing. Roddenberry and Herb Solow, the head of TV production at the studio, asked my permission to run some video tape on me to be studied in the projection room. I agreed and was led onto the stage of "I Love Lucy." Fortunately the audience had not arrived for the day's taping. But the crew, fifteen or twenty craftsmen, was asked to light me and run some footage.

Had the makeup been complete, the wardrobe present, the character fully realized, it would have been difficult. Under the circumstances, dressed in casual street clothes, with a crude pair of pointed ears, and in the context of the "I Love Lucy" set, it was, to say the least, painful. And yet, here it was—the beginning of the public exposure of the extraterrestrial.

I found myself making mental notes, storing away emotional memories that might someday be useful in the role. The feeling of being alien, almost to the point of being ridiculous, of knowing that members of cast and crew and onlookers would be composing clever lines of dialogue to exchange after I had gone—these were the real seeds from which the emotional structure of Spock would grow.

I was moving into the world of an extraterrestrial. Already I could feel myself building defenses, attempting to elevate my thinking above and beyond a concern for the opinion of *mere humans*. I was of another realm, and they could think what they would. Several weeks later I had a similar experience, on the first day of shooting the pilot film. Fully wardrobed in the uniform prescribed for Starfleet and in full makeup, I stepped onto the sound stage at the studio for the first time. This time, at least in a physical sense, the character was complete.

My agent was there to greet me. With him was a lovely female client. Her reaction was startling. She was open and obvious in her interest, and gave me my first exposure to the generous and gratifying female reaction that was to come, synthesized in Isaac Asimov's description of Spock as a "security blanket with sexual overtones."

So it began. I went to work and played the scenes. I groped—trying to learn to walk, talk, and function as an alien. I put out the sounds and motions, and then watched—recording the feedback of my fellow actors, crew personnel, and visitors.

For a long time it was difficult. The total understanding of the character would only be found in the total context. This rigid, pointed-eared creature was only a visual gimmick until perceived as part of a whole—the particular story we were filming, the ship, the crew, the antagonists, the entire creation of another world, another time. Taken alone, in bits and pieces, out of context, it was still dangerously close to a joke.

Nevertheless I gradually began to feel more comfortable. I could understand my place, my function in the stories, my relationship to the other characters. I began to develop a sense of dignity, a pride in being different and unique.

Above all I began to study human behavior from an alien point of view. I began to enjoy the Vulcan position: "These humans are interesting, at times a sad lot, at times foolish, but interesting and worthy of study."

The scripts, particularly the character of Dr. McCoy, the humanist, offered opportunities to deal with the human need to see everything and others only in relation to themselves. "Anybody who isn't like us is strange. Anybody who doesn't want to be like us is a fool."

I was becoming alienated, and I didn't realize it. My attitude toward the humans around me became quite paternal. In some respects I assumed the position of teacher or role model. My hope was that we could reduce inefficiency and silly emotionalism if I set examples through my higher standards of discipline and precision.

Nature abhors a vacuum, they say. I showed little or no emotional response, so my coworkers and associates projected responses for me. For example, this comment by a coworker was passed along to me by a friend at the studio: "I see [in Nimoy] a growing image of a shrewd, ambition-dominated man, probing, waiting *with emotions and feelings masked,* ready to leap at the right moment and send others broken and reeling."

At the outset Nimoy-the-actor felt protective of the character, much as a parent is protective of a child. Certainly my ego was involved and was bruised. But it seemed important to help the character to come to life. When he did, he protected the actor. He became an ever-present friend who could be called upon as an ally in adverse circumstances. Nimoy could submerge himself and let the formidable Spock take over.

Eventually, in September 1966, the show went on the air. The reaction was immediate, and the popularity of the show grew at an astounding rate. The magic of Spock became apparent very quickly.

I was mobbed at personal appearances, and security measures were necessary to get me into and out of crowded situations. Beneath all the excitement I sensed a real desire to communicate on the part of the fans. There was more to their reaction than the typical attention given a TV personality, particularly in the case of young people through college age.

For a long time I have been aware that many people would rather meet and talk to Mr. Spock than Leonard Nimoy. Students who come to hear and see Leonard Nimoy arrive with mixed hopes and expectations. To what extent is he like Spock—in appearance, in manner, and in thought? Will they be able to experience the Spock presence, or is this to be a human-to-human contact? They will accept Nimoy respectfully and attentively. They are curious to hear what he has to say about himself, his thoughts and interests, and so on. There is probably an ongoing comparison of Nimoy with Spock, a search for those human elements that Nimoy must have replaced with Vulcan characteristics.

Occasionally—by choice or accident—Nimoy steps momentarily into the Spock character. The response is a wave of joyous recognition. It is as if Nimoy had stripped away the human facade and given the audience a glimpse of the *real* person, the Vulcan disguised as a human. Followers of the "Star Trek" series are so attuned to the character traits of Spock that they respond to even the most subtle hints that they are in the presence of this alien.

Following the opening applause, I can open a lecture with a simple icebreaker: "You obviously watch a lot of TV." This is a neutral comment, establishing a common meeting ground. On the other hand, the flat delivery of a Spock line, such as, "You're a very emotional group of humans," immediately releases the charisma of the ETI and gets a strong, positive response. They seem to be saying, "Talk Spock."

During a question-and-answer period, the audience delights in seeing or hearing the typical Vulcan response to some of the questions.

FEMALE QUESTIONER: How old are you?
SPOCK: [*straight-faced*] What did you have in mind? [*laughter*] . . . The question obviously reflects your need to judge the possible relationship between myself and other beings, possibly even yourself. This represents an illogical human compulsion on your part, based on certain insecurities and a fascination with linear time. It would be better if you humans could rid yourselves of this fixation.

The audience response would seem to suggest a delight in seeing the questioner or even the human race "put down," particularly if the audience recognizes the question as being childish or fanlike in nature. The position seems to be: Spock should not descend to the level of the questioner. He is not so foolish as to give serious answers to silly or unimportant questions. In fact, he elevates our thinking by pointing out the foolishness that is inherent in the question.

Among the individuals in the audience there is also a feeling of relief, as if they were thinking, "That was a silly question, and I'm glad Spock agrees with me."

College audiences are particularly responsive to Spock. If I were to select a committee to make the initial contact with an ETI arriving on earth, I would certainly include college students. Their interest, their sympathy, and their ability to empathize with new and perhaps exotic ideas would be invaluable. Speaking for Mr. Spock, I can definitely say that they make him feel very welcome.

Why? Why does a generation raised under the influence of Dr. Spock relate to Mr. Spock? Is Spock embraced and welcomed because mankind is troubled by the march of events? Here is an ETI of superior intelligence and abilities, capable of making difficult decisions free of ego and pressure and emotional need, dealing (supposedly) only with the facts in each case and arriving at the logical conclusions. The period in which Spock arrived was one of polarization over major political and social issues. The war in Vietnam, the drug culture, the Black revolution, assassinations, and so on.

Perhaps Spock represents a wise father figure to whom humans can turn for solutions to thorny problems. In many cases humans are torn between doing the right thing and doing the expedient thing. "I know what's right, but if I do that, what will my neighbors say? What effect will it have on my family? What will it cost me?" Spock seems to be free of these tensions. In the Vulcan culture one simply does what is right.

We must not overlook the fact that we know that Spock is part human and therefore suspect him of being compassionate, even a humanist at heart. Thus we feel safe in placing our fate in his hands. Certainly he would never make a decision that, though logical, would be antihuman. Logic alone might someday dictate the extermination of millions of innocent people in order to relieve overpopulation, food shortages, and ecological problems. In that case we could turn to Spock-the-scientist and know that he would find brilliant solutions to mankind's needs.

So this particular ETI or type of ETI is superior in his decision-making abilities and in his scientific knowledge. But we trust that he will apply these superior assets for our benefit.

The teen-ager coping with the fiercely complex problems of adolescence often feels very much alone. His friends or peers are understanding, but they too are faced with the same problems and have no solutions. They can only commiserate. Parents claim to have the answers, but they are short on patience and understanding. Then too, the child-parent competition makes it difficult for the child to accept the parent's solutions. They seem old-fashioned or, at best, out of touch with contemporary attitudes. Too often the parent's advice is couched in emotional terms.

Spock easily resolves any dilemma. He has superior insight. He quickly understands the nature of the problem. He has studied the human race, and he is a pure authority on the subject. He has no ax to grind because he is so totally secure. He can certainly not be accused of being old-fashioned. He is future. He can be compassionate in his judgment and dispassionate in his help. To the young female, he poses no sexual threat. Spock is asexual. She need not fear that he will take advantage of her. In fact, quite the contrary, he probably becomes an object of sexual fantasy because he is at once safe and challenging.

I was doing some homework one day. Without being conscious of doing it, I was quietly singing a song by Jacques Brel:

> If we only have love
> We can reach those in pain
> We can heal all our wounds
> We can use our own names.
>
> If we only have love
> We can melt all the guns
> And then give a new world
> To our daughters and sons.

I was startled by Spock's voice from behind me: "What is that song?"
NIMOY: Well, it's a poetic idea. It suggests that we could all get along with each other if there were a common love.
SPOCK: Do you believe in this concept as a possibility?
NIMOY: I see nothing wrong with believing that mankind could someday find real brotherhood.
SPOCK: There is no need to be defensive.
NIMOY: Spock, there are some things that we humans care about that may seem strange, even ridiculous, to you. What's wrong with a little hopeful poetry?
SPOCK: It is not the poetry I am curious about. I understand it. It scans fairly well, and some of the images are quite lovely. Poetry has an important place in Vulcan literature.
NIMOY: Then what's your point?
SPOCK: The poet suggests that love can stand alone as an emotional entity.
NIMOY: Where does he do that?
SPOCK: "If we only have love, we can melt all the guns."
NIMOY: What he means is that if people loved each other there wouldn't be any more wars.

SPOCK: Illogical.

NIMOY: Why?

SPOCK: War is the result of a breakdown in reason. It is an emotional confrontation. To eliminate war, it would be necessary to eliminate emotion. Love, being an emotion, would thereby be eliminated as well.

NIMOY: Are you saying that we can't have love *and* reason?

SPOCK: Most human literature would suggest that reason disappears in the presence of love.

NIMOY: Under the influence of a drug you once fell in love.

SPOCK: True.

NIMOY: When it was over I heard you say, "I was happy for the first time in my life."

SPOCK: Your typically human assumption is that I had till then lived a deprived life, but, because of that moment, I had at least had an insight into a blissful state.

NIMOY: Isn't that what you meant?

SPOCK: Not necessarily. My statement can also be interpreted to mean, I have had an experience peculiar to the human race.

NIMOY: That's not what it sounded like when you said it.

SPOCK: That is a matter of individual interpretation.

NIMOY: Didn't you like the experience.

SPOCK: I can understand that you would *want* me to like it.

NIMOY: Why?

SPOCK: Because you want me to be like you, to value the things you value and to give support and credence to your lifestyle.

NIMOY: [*losing patience*] Look, Spock, all of this is very interesting. But you did say you were *happy*. By definition, that's a pleasant emotional experience.

SPOCK: I know this is difficult for you, but please try to understand. I said, "I was happy for the first time in my life" in the same sense that a visitor to America might say, "I had a hamburger for the first time in my life." Of course most Americans would like to hear him add, " . . . and I loved it."

NIMOY: Spock, you're impossible.

SPOCK: Are you angry with me?

Spock is a good man to have around. He is brilliant, dignified, loyal, and cool. In a crisis where humans stumble, Spock functions with logical efficiency. He is extremely honest and incorruptible. He seems to be compassionate, although he would deny it because compassion borders on emotionalism.

There is a suspicion that Spock has secret knowledge that could be helpful to a troubled humanity. I have been contacted by "metaphysical" organizations. One told me I was chosen for the Spock role because I was a carrier of ideas, of concepts of which I myself might not be aware, and that my real function in the series was to prepare mankind for the future and to relieve public tensions about phenomena that were bound to come. This particular group communed in the Nevada desert and had personal contact with extraterrestrials who visited them and took members of the group on short rides in their spacecraft. They could reveal themselves to me, they said, because I would be capable of accepting and understanding. However, they felt the public would be hostile and needed further education through roles like mine.

Spock is a safe and comforting answer to the extraterrestrial question. Many people have gone to great lengths in writing to me about the effect he has had on their lives. They often tell me, movingly, that he demonstrates a dignified way for the individual to function in what, for many, is a hostile society.

If there are extraterrestrials and if we are to encounter them or be visited by them, most people would be relieved if they turned out to be like Spock.

It is easy to forget that Spock was created by humans. And if he does in fact represent elevated concepts of life, its value and its meaning, then those elevated concepts must be credited to the human race. Personally I find it an enormously exciting challenge to try to live up to the broader vision and deeper perception that I helped build into this fictional character.

Tennis players tell me they prefer to play against a superior opponent. It sharpens their game.

NIMOY: Spock, how does it feel to be popular?
SPOCK: I do not have feelings.
NIMOY: I'm sorry. I didn't mean to offend you.
SPOCK: I am not offended. I understand your tendency to judge me by your human standards. It would, however, facilitate matters if you would refrain from doing so.
NIMOY: I'll try. . . . Are you aware that you are popular?

SPOCK: I am aware of a certain public interest that exists.

NIMOY: People *like* you. Do you care about that?

SPOCK: Should I?

NIMOY: Well, I realize that it's not something you would pursue, but being liked is a nice thing.

SPOCK: Is it useful?

NIMOY: I suppose not.... But I don't think it's destructive.

SPOCK: Many of your political leaders have achieved much popularity and left behind much destruction.

NIMOY: Then, would you rather not be popular?

SPOCK: To be concerned one way or the other is a waste of energy. And popularity does put one in strange company.

NIMOY: How do you mean?

SPOCK: In your culture popularity may be achieved by bizarre beings and in strange ways. One can achieve popularity by appearing nude in one of your magazines. Certain animals have become popular through weekly exposure in your television dramas. Would it not be better to honor real achievement?

NIMOY: Can't popularity and achievement go together?

SPOCK: Victor Hugo said popularity is "the very crumbs of greatness."

NIMOY: That sounds pompous.

SPOCK: Possibly. But after all, Mr. Hugo was only human.

NIMOY: To be more accurate, Spock, Hugo was only *a* human.

SPOCK: True.

NIMOY: And though there might be some validity to his statement, there are many humans who have expressed diverse ideas on this and other subjects.

SPOCK: True.

NIMOY: Would you be satisfied if I, as a human, took any random comment from Vulcan literature as representative of all Vulcan thought on the subject?

SPOCK: To do so would be to deny the usefulness of—

NIMOY: And if you are popular among humans, doesn't that say something positive about the human ability to value a culture and a lifestyle alien to its own?

SPOCK: That does seem logical.

NIMOY: Mr. Spock, coming from you I consider that a great compliment. From the bottom of my emotional heart, I thank you!

How might our Western tribal religions stand up to an encounter with extraterrestrial life-forms, whose very existence may threaten cherished religious beliefs? Using a television documentary format, William Hamilton explores some possible reactions.

For whatever labels are worth (and they are of great worth only to dogmatists and categorists—the pun is Korzybski's), Bill Hamilton was labeled a "radical theologian" during the 1960s, when his pioneering publication *The New Essence of Christianity* (1960) identified him as an articulate spokesman for the "death of God" movement in Christian theology.

His many writings since then include *Radical Theology and the Death of God* (coauthor, 1966) and *On Taking God out of the Dictionary* (1974). An excellent summary of the death-of-God position can be found in the commissioned article that Hamilton wrote for *Playboy*, "The Death of God" (August 1966).

Hamilton was also the writer for, host of, and (occasionally) actor in the CBS public-affairs series "Look Up and Live!" He produced a variety of television and radio plays and documentaries on religious issues.

Dr. Hamilton is at present Dean of the College of Arts and Letters at Portland State University in Oregon.

"I see no objection to saying that God may have created other worlds than this one. I see no objection to saying that a savior may have visited other places in other times. I see no objection, but neither do I see a point. . . ."

William Hamilton

The Discovery of Extraterrestrial Intelligence: A Religious Response

What follows is the transcript of a television special, produced by CBS News and aired on September 23, 1989.

THE PRESUMPTION OF MEDIOCRITY

A series of special reports on the
significance of the recent discovery of
extraterrestrial intelligence
[Part V]
The Religious Meaning

[*Sound effect: The tape of the original message is heard, a series of regular radio pulses in patterned order, with regular intervals between.*]

ANNOUNCER: On June 15, just before midnight, the radio signal you are now listening to was received at the massive radio installation for

99

galactic communication in Parkes, Australia. Scientists quickly determined that this signal emanated from a previously unknown planet orbiting Alpha Centauri, a star approximately four light-years away. Thus the message we received originated approximately four years ago. Since June 15, further messages have been received and the decoding process has begun. For over thirty years scientists all over the world have been training their radio receivers on countless stars in our galaxy, convinced that extraterrestrial intelligence exists and also convinced that it is seeking to communicate. We do not know, of course, whether other civilizations have received such messages, but we do know that we have received this message. The probability of extraterrestrial intelligence has become a reality, and man is entering a new era, for good or ill. CBS News is presenting a series of special programs on this event entitled "The Presumption of Mediocrity." This is our fifth program in the series: "Extraterrestrial Intelligence: The Religious Meaning."

REPORTER: In earlier reports we examined the scientific background of this momentous event. We showed you how the signal was received and how the original message of pulses and intervals was decoded into information, taking a pictorial form. We noted that for many years scientists have assumed, often without hard evidence, that there were many planets in our galaxy that could contain highly developed technological civilizations. Such advanced civilizations may number in the millions. We now know of one.

We already possess a little information about the planet originating the signal. We know its size, the length of its day and year, its distance from its star, Alpha Centauri. We are beginning to learn something of the planet's surface temperature. And just last week the longitude and latitude on the planet of the transmitter was established.

Tonight we wish to turn from these dramatic scientific findings to the problem of our response. What does this event mean for us, as humans and as Americans? How are we to respond—with fear or hope? What, if any, is the religious meaning of this event?

Some years ago, when our astronauts began to return from the moon, it became clear that the effect of that earlier technological achievement in the realm of imagination or spirit was fully as significant as the effect on science and technology itself. Tonight we wish to look at the spiritual effect of this new discovery.

One of the early astronauts, some twenty years ago, said of his trip to the moon: "You develop an instant global consciousness, a people orientation, an intense dissatisfaction with the state of the world, and a compulsion to do something about it. . . . Something happens to you out there." What will happen to us "down here"? Early in the 1970s a distinguished astronomer who had thought deeply about the problem of extraterrestrial intelligence wrote: "I think no serious student of the subject questions its significance—both for science and for the deepest philosophical and human questions. In a very real sense this search for extraterrestrial intelligence is a search for a cosmic context for mankind, a search for who we are, where we have come from, and what possibilities there are for our future—in a universe vaster both in extent and duration than our forefathers ever dreamed of." As we have tried to assemble for you a portrait of America's initial religious response to the discovery of extraterrestrial intelligence, we have found other expressions than those of optimism. We have found fear as well as hope, deep defensiveness about old ways as well as predictions of a new religious basis for cosmic unity.

The religious climate in the America that first received these signals is not an easy one to describe. We put this question to a number of observers of the religious scene: "How would you describe the status of religion in America prior to June 15th of this year?"

[*A series of film clips follows:*]

(1) "I wouldn't dare give a confident answer to that. In some ways we have never been more secular, more self-satisfied and privatistic, more devoted to pleasure and consumption. Churches and synagogues have steadily lost their constituencies and have become the refuges of the worried and the aging. The young still have their moments of religious concern, largely unrelated to the institutions of religion, but these moments tend to be short-lived, as they settle into their roles in the consumer society. They find, as their parents found before them, that consumption absorbs all their spiritual energy."

(2) "Conservative religious movements, built upon guilt and promising absolute certainty, have been flourishing for decades, and I suspect they'll continue to flourish in this new climate. Religion as a social force is about dead, just as people's confidence in politics is dead. If we are religious at all, we are so in a deeply private sense—religion as a private trip to private security. The alliance between conservative religion and right-wing politics has been running this country for a decade, and I suppose it will continue to do so."

(3) "I'm tempted to give a Marxist answer to your question, though I'm no Marxist. If you've got a job, then your life is an exhausting dialectic

of compulsive work that you hate and compulsive play that is hardly more enjoyable. You have no time for religious longings. If you are unemployed or unemployable, then some religion may still serve as a last resort."

(4) "Traditional religions are dying because Americans don't worry about sin and guilt any more. Everything is permitted, and everything is done by someone. I'm not even sure that anyone worries about death in such a way that religion might be an answer. If we do worry about it, then exercise and diet are our responses, not God and immortality. What we worry about is cancer, heart attack, obesity, growing old, poverty, money, sexual adequacy. And Christianity and Judaism, at least, don't have many promises to make along those lines."

REPORTER: So much for the state of religion in the country where the signals were received a little over three months ago. What has happened since? Are we returning to God, or to anything equivalent? To get some feeling for an answer to this question, we took our cameras last week to what is probably an atypical Protestant church here in Washington, D.C.—St. James Episcopal Church, across the street from the White House. It is atypical in many senses: it is relatively successful, as churches go; its congregation is not a homogeneous one, but is both black and white; upper-class, lower-class and in between; educated and not so educated. The congregation includes many government employees and a number of scientists working in the capital. The rector, Dr. James W. Temple, preached last Sunday on the problem of a religious response to the new discovery. His sermon seemed to us so suggestive and stimulating that we decided to bring it to your attention. We are also including a portion of the lively discussion that followed.

RECTOR: I don't suppose I need to apologize for preaching a sermon this morning. It's been several years since I've done such a thing, but it can be justified. The radio signals that were picked up early this summer are all we've been talking about all summer, and this is our first regular service in the fall. I myself have been talking and listening to literally hundreds of people, in official meetings here and in New York, and unofficially all over the place. I find I need to get my own thinking clear on the meaning of this important event, and some of you may be helped by overhearing me and, if need be, taking issue with me. None of us are experts here; we're all neophytes. I have a biblical text to propose. It is the obvious one, from Genesis 1:1: "In the beginning, God created the heavens . . . "

Extraterrestrial intelligence—for many years a possibility, for some years a probability—is now a reality with which you and I have to live.

What has been, I ask myself, the initial effect of this discovery on my faith as a Christian man?

My answer is curious, and it surprised me. I discover that my Christian faith is neither much of a help nor much of a hindrance. I don't feel that the discovery makes my Christian allegiance any easier, or any harder. I know the newspapers have been reporting other reactions. We're being told that Christianity is suddenly outmoded. Perhaps, but remember that we have never legitimately defended ourselves on the grounds that we are fashionable or up-to-date with the latest scientific discoveries. We're also being told that the Bible has now been proven literally true, that these extraterrestrial intelligences are really the angels of Christian legend, and so on and so forth. Well, let's wait and see before we follow either of these directions. Meanwhile I think we should try to avoid an anxious defensiveness. It's not important for us to be found correct, but to be found humane, just, and loving.

We have received, I suspect, something of a shock to our egotism. (As a Christian I am predisposed to approve of shocks to egotism.) Some years ago, shortly before he died, the Swiss psychologist Carl Gustav Jung predicted that the discovery of extraterrestrial intelligence would mean that the "reins would be torn from our hands and we would, as a tearful old medicine man once said to me, find ourselves 'without dreams,' that is, we would find our intellectual and spiritual aspirations so outmoded as to leave us completely paralyzed."

Are we indeed now without dreams? Some of us, perhaps. Those of us who have been making our religious way through life by turning our backs on the new, with dogmatic and authoritarian assurance, with a conviction of our privileged and unique character—such individuals may be now receiving a shock from which they will not recover. But we have received shocks to our egocentricity before. Sigmund Freud pointed out that when Copernicus established that earth was not our cosmological center, and when Darwin established our kinship to the animals, and when he himself pointed to the centrality of sexuality to our behavior—cosmic egocentricity received decisive blows indeed. The news of this summer has brought no less a shock than these earlier ones, but perhaps no greater either. And the biblical message about the sin of pride may have made all these shocks somewhat less surprising. Perhaps only the chauvinist and the dogmatist are truly restless tonight.

America, as it moves into the last decade of this millennium, is a troubled land, probably a sick land. We have suffered under a strange

kind of benign political dictatorship for more than ten years, as I hardly need to remind you who work in this city. Our people seem largely bored, lazy, with no intellectual or moral passion save that demanded by the search for consumption and pleasure. Our classical enemies are still with us—poverty, disease, overpopulation, despair—and we have despaired of conquering any of them. Let the government ineffectually play, we cry, so long as they leave us alone.

Will the discovery of this new civilization, offering us a chance to communicate, be able to cure our moral torpor? Even as we have learned to despise the damage our familiar arrogant technologies have done to us, may not this new technological event initiate a call to a new adventure, with a new world to learn from? I do not know; none of us do. I suspect things will become, in our lifetimes, either much better or much worse, a new Renaissance or a new Dark Age. Both are possible, and I am going to suggest that the actual direction does not depend on the gods or on the scientists but on all of us together. Our response will determine our future.

So much for introduction. I have two main points I wish to make; the first is moral, the second religious and theological.

One fact emerging from our discovery of extraterrestrial intelligence is the basis of both our fear and our hope: this new civilization, suddenly burned into our consciousness, must be assumed to be more advanced than ours. This is the meaning of the phrase you've seen so much in recent weeks, "the presumption of mediocrity." It must be presumed that we, compared to them, are the mediocre ones. This is only partly because it is more difficult an achievement to send an interplanetary signal than to receive one. The presumption of mediocrity is based on a more sophisticated argument. It goes like this, as far as I understand it: Human life or, more exactly, humanlike life has been on the earth for about one-tenth of 1 percent of earth's history. Civilization, in its broadest sense, has been in existence for about one-millionth of the earth's lifetime, while technological civilization has been with us for only one-billionth of the earth's geological time. Against such figures, we are bound to assume that this newly discovered civilization is both older and more advanced than we. Now, is this good news or bad?

If it is more advanced, then it is certainly stronger. Should we fear conquest? Apart from the question about the probability of future physical contact between the two civilizations—and the answers to that question today vary from a few decades to never—conquest seems unlikely at this early stage of speculation. If they are indeed more advanced, it appears to follow that we will pose no threat to them. It is unlikely that they will want us for slaves since presumably their advanced technologies will have solved the problem of burdensome labor in some mechanical way. If not for slaves, then what? They could conceivably look upon us as we look upon lower animals on the evolutionary scale—as amusing pets or means of public entertainment. We might be to them as seals and dolphins and circus animals are to us. Or they might wish to destroy us for mere amusement, much as sportsmen treat wild game. Or, moving out of fancy and a little closer to probability, perhaps their main desire will be to evangelize and convert. You might want to argue against this on the grounds that only a fearful and insecure ideology is likely to be concerned with proselytism. The evangelical impulse, it can be argued, is found only when an insecure group tries to assure itself of a desired superiority that it actually doubts. Our very mediocrity may give us protection from their missionary zeal.

Up to now the radio pulses have not been translated into language, only pictures. So we do not yet have any information about their moral attitude or intention. It appears now that what they want from us is precisely what we want from them—information. Information for its own sake, and information that can be used, and mutually exchanged, to improve our lives and our spirits. There is another basis for optimism. Many in the past have wondered about the stability of our own civilization. Is it inevitable that a technological civilization germinates its own seeds of destruction that come to full growth in some finite time interval after, let us say, the discovery of radio communication? Philosophers of history have been predicting our doom for generations. But look, if this new technological civilization, far in advance of us—even thousands or millions of years in advance—is surviving, then predictions of our imminent doom begin to ring hollow. The apocalyptists, secular and religious, are presumably wrong. We can still destroy ourselves, but it will be our free choice that does it, not a malevolent technological providence.

On balance my hopes outweigh my fears, which recalls the ancient wisdom that if hopes may be dupes, fears may be liars. We must begin to let our imaginations play on what this information explosion might bring us. Think what a technological civilization thousands or millions of years in advance of ours might be able to teach us—the cure for disease, better methods for growing and distributing food, insights into the realms of beauty, wisdom, spirit. But can we manage the humility it takes to learn from superior wisdom? Easily, I suppose, when it comes to the cure of cancer. But when it comes to politics and religion and perhaps even human sexuality—to be willing to learn and to grow and to change in these areas requires a kind of moral sense, a tolerance and a humility of a high order. Are we, or can we become, a people moral enough to learn? "Unless you become as a little child, you shall not enter the Kingdom of God." If it is more blessed to give than to receive, it is surely more difficult to receive than to give.

I must confess that such moral reflections, set in motion by the new discovery, interest me more and even trouble me more than the obvious questions about religion and theology that the gossip mills have been chattering about over the summer. There are religious and theological questions to be faced. Can we still believe in God? Can we still be Christians or Jews or atheists? Are there any atheists in intergalactic foxholes? Is it not possible that such questions augur already a kind of grim defensiveness quite the opposite of the resourceful humility I have been arguing for?

But we must turn, nonetheless, to these religious and theological matters. (I must admit to some skittishness, for I am perfectly aware that I am more likely to give offense here than in my earlier remarks. Who could possibly be offended by defenses of humility, particularly from a pulpit? That's what you're supposed to hear from a pulpit.) The ultimate thing, the bottom line in religion and in religion's theological explications, is that we learn to become as lovingly and responsibly human as we can be. If our religions and theologies, our rituals and creeds, our priests—masculine and feminine, straight and gay—if these can help us become so, they have their validation. If they stand in the way of our full humanity, they must be set aside. If we should find, as we face the infinitely exciting new adventures proposed to us by this discovery, that some or all of these things are making us more self-absorbed, less resilient morally, less open and loving, then we are not only permitted, we are obliged, to put them to death. And anything, even a technological discovery, that helps us put them to death can only be a good thing for us.

There has, of course, been a great outpouring of official religious utterance over the summer, most of it designed to assure believers that nothing in the discovery of extraterrestrial intelligence in itself disproves *their* religion, whatever it might be: God might just as easily have created that other planet as well as this one. Recall our text: God created *the heavens*. The Bible says so. We're home free. Christ might have visited, perhaps in some other form, perhaps in some other time, this or that other inhabited planet, and brought to others the reality of the same God he once brought to us. A generation ago the British theologian and science-fiction writer C. S. Lewis found the form of Christ all over the galaxy. Can't we do the same?

Is this really the way we want to go? I for one do not have the slightest interest in exploiting the discovery of extraterrestrial intelligence for the purpose of defending Christian truth. I see no objection to saying that God may have created other worlds than this one. I see no objection to saying that a savior may have visited other places in other times. I see no objection, but neither do I see a point.

I think we are tempted to move in this defensive direction only if we are deeply convinced, or deeply need to be convinced, that Christianity is the only true faith and that it is thus true for all times, all places, and all galaxies.

But be careful before you are persuaded by an argument against Christianity's absolute truth. If we say that Christianity is appropriate, or even somehow true, for only one time, one space, one planet, then is this relativized Christianity really Christianity at all? Expand the Christian God into the architect of a new cosmic consciousness. He may still be God, but is he Christian? Take away from Jesus Christ the unique, once-for-all Incarnation, death, and resurrection and throw him into the skies. You may still speak of a savior, but of one who is no longer the Christian Lord. The conservatives among us, anxiously defensive, insisting on Christianity's absoluteness and uniqueness, may be religiously wrong in some ways, but they may be right in their insistence that Christianity stands or falls on that absolute claim.

If I am to keep religiously open, if everything I find I hold to must now be deemed corrigible, am I not on the slippery slope (or is it open road?) from the Christian to the post-Christian world? Do we really need the idea of God to give shape to our emerging consciousness? Isn't this new consciousness, just because it is no longer based on our limited planetary experience, going to assume a post-Christian—and therefore a non-Christian—substance? All of our old religions, Western and Eastern alike, have been deemed true because they enable us to live and love effectively in our concrete experience of time and space. But that basic experience is in radical disarray, and each of the old religions to that extent may have become false.

Do we need God for this new work? Or Nirvana, or Brahman, or Torah? I think not. What we do need is a way of pointing to the world of mystery we live in, a continuing criticism of our pride and sensuality, a way to humility and love. We need something that can render our familiar lives, and the new and unfamiliar lives we are moving toward, more open and human. This is why we can still readily call ourselves religious men and women. But perhaps we must prepare ourselves—just because of what the Christian religion has meant to us in the past—to do without the old and familiar ideas and symbols. For the sake of Christ, we may need to go beyond him. We may need to de-Christianize our imaginations and our lives so that in the new context we can live the life of discipleship to which Jesus once called us.

This will be the consequence if, in the face of this discovery of extraterrestrial life, we stop busily defending ourselves and assuring ourselves that our gods are intact and our faiths unshaken. We ought to be shaken. Let me put it strangely. God may want us to go beyond Christianity, and even beyond him. "Before God, we are without God," as Dietrich Bonhoeffer wrote many years ago. If we retreat, in the face of this arresting challenge to our imaginations and spirit, to our old and comfortable convictions, we will find ourselves curiously disloyal to the very meaning those convictions once so splendidly bore. Let us make the venture. Let us exact a careful look at our inherited spiritual and intellectual furniture and get rid of everything—even God—that keeps us from being open and attentive to one another. Let us listen carefully and put final confidence only in ourselves.

Many years ago, when I was young and liberated, a rock group called the Beatles meant as much to me as anything in my life. In recent days I have found myself recalling a fragment from one of their old songs, called "Let It Be." It suggests the inner feeling, the religious feeling, that I wish to call myself to and perhaps to call you to as well:

> When I find myself in times of trouble,
> Mother Mary comes to me,
> Speaking words of wisdom,
> "Let it be."

REPORTER: This sermon was delivered last week by Dr. James W. Temple, rector of St. James Episcopal Church, Washington, D.C. It may or may not have been representative and typical of other religious discussions in other religious traditions across the country, inside or outside buildings. It seemed to us to be at least suggestive. After the service our camera crew joined Dr. Temple and some of his parishioners in the church lounge, and we want to bring to you a portion of the lively discussion that took place. Clear answers to the religious questions raised by the new discovery will be a long time in coming. Even to raise the right questions is a demanding task.

FIRST QUESTIONER [*a middle-aged man, probably a vestryman, clearly irritated by what he has just heard*]: You apparently have no qualms in asking us to give up our faith just because of some damn radio signals. Are you really planning to throw everything out of the window—God, Christ, even immortality?

RECTOR: I am not really asking you to do anything. I'm thinking out loud, and my thought has surprised even me by its direction. On immortality, Dave, I really think that the traditional beliefs are worthless.

YOUNG WOMAN: The "people" at the other end of those signals may not even be people. That civilization may have far outgrown its humanoid phase. If it has, it's possible they have no experience of death at all.

RECTOR: I want at least to be ready for the possibility that this other form of life may not at all experience death as we do—as a threat and enemy to be feared and perhaps, at the end, uneasily accepted. Their life may be more mechanical than human; they may wear out or run down rather than die. Or death may be something they choose, as we choose a lover, rather than something that comes upon them. Or their medical technologies may have eliminated much of what we mean by immortality. I don't want anything in my traditional background to keep me from learning from a civilization with an experience totally different from my own. God and Christ and immortality are for me on the bargaining table. We might even come to learn that immortality isn't a religious term at all, but, say, a technological achievement.

SECOND QUESTIONER: I'm not quite clear, Rector, just why you feel that this new discovery may have called Christianity into question in some fundamental way. After all, the church did survive the egocentric shocks of Copernicus, Darwin, and even Freud.

YOUNG MAN: I think I know part of the answer to that. We scientists have always held, more as a hypothesis than as a proven theorem, that nothing in nature—not even the appearance of life as on this planet—takes place only once. We tend to be skeptical about unrepeatable events; the scientific method itself is closely connected to repeatability. This hypothesis has been strikingly confirmed by the discovery of another intelligent planet, and now we're even more uneasy than ever with events—

RECTOR: Like Incarnation and resurrection—

YOUNG MAN: Yes. With events that are supposed to have happened once and for all.

RECTOR: That is why I argued in my sermon that Incarnation becomes impossibly distorted and finally dechristianized when it is spread all over the galaxy. After all, science and theology have been on this methodological collision course ever since the seventeenth century. This is just the last and, I suspect, the decisive battle in that war.

YOUNG WOMAN: It never occurred to me that we might be in touch with something that isn't really human at all.

RECTOR: More likely to be superhuman (is *transhuman* a word?) than subhuman or inhuman. I think we all naturally assume that the human state must be the highest state of the evolutionary process, because it is so for us. But this other civilization, perhaps a million years in advance of ours, may have achieved something better, certainly something different. Our ultimate problems are all human ones: sin, guilt, fear; and our goals are human: justice, humility, love. We cannot conceive of the problems and goals of the posthuman stage of evolution.

YOUNG WOMAN: Unless we've been reading science fiction.

RECTOR: Yes. I suspect we'll have some sort of answer to this question about the form of the posthuman consciousness fairly soon. Messages are being received almost continually and being decoded and interpreted right along. We may well be able to find out whether our humanity is indeed advanced or, in the new context, quite impossibly primitive. Last week I came across a beautiful quotation from a distinguished scientist from about twenty years ago, Loren Eiseley, in *The Immense Journey*. I decided not to use it in the sermon, but it's here in my notes somewhere. . . . "But nowhere in all space or on a thousand worlds will there be men to share our loneliness. There may be wisdom; there may be power; somewhere across space great instruments, handled by strange, manipulative organs, may stare vainly at our floating cloud wrack, their owners yearning as we yearn. Nevertheless, in the nature of life and in the principles of evolution we have had our answer. Of men elsewhere, and beyond, there will be none forever."

ANOTHER QUESTIONER: I would certainly like to know what other religious people around the world are thinking about.

RECTOR: We're beginning to pick up some bits and pieces of information. We're all a little threatened and a bit defensive. That's understandable, and it's bad only if it's the only reaction. Buddhists may have the easiest task of adjusting. They have never had any difficulty with the idea of worlds beyond this one, mainly because they have never been dependent, as Jews and Christians are, on a particular piece of historical time as the locus of an historical revelation. Hindus can talk readily about the transmigration of souls into other bodies or spiritual states, but as far as I know they have never envisioned other inhabited worlds. The Confucians have always emphasized this world and man's behavior in it. In this early phase of the discussion, we must all be very much alike—a little worried, a little defensive, a little excited.

ANOTHER QUESTIONER: You seem to think all this may take us into some sort of post-Christian realm, but not, apparently, postreligious. You don't think, I take it, that this discovery will turn us toward a more secular direction. You did say "away from God," but you also said "toward religion."

RECTOR: It just occurred to me that perhaps it is a bit arrogant of us to talk quite so much about "discovery," as if *we* had done something sensational. They're the ones who have discovered us. You're right. I don't expect any significant move toward secularism, toward religionlessness, to use an ugly word. I suspect the direction will be the opposite, away from secularism and from Christianity to some new form or forms of religion. They may be lovely; they may be horrible; they may be both.

ANOTHER QUESTIONER: What do you mean when you say "religion"?

RECTOR: That's a fair question, and a tough one. Religion has to do, I think, with two things. First, religion is born of the fact that we are not in absolute control of our lives or our destinies. We have a sense of something more, something beyond, something we can't grasp. As humans we always stand before mystery. Wonder, reverence, awe—these are some of the names we give to our responses to this mystery. And as I mentioned earlier, religion permits but does not require the idea of God to give linguistic expression to this sense of mystery. Religion without God—or as I would put it, religion beyond God—is a curious idea for Christians, perhaps, though familiar enough to Buddhists.

Second, religion has to do with what you value—what you give worth to, what you put at your center and what at your periphery, and how you order and rank and justify your values. When we think carefully and undefensively about the presence of this new center of extraterrestrial intelligence and when we define religion as I have tried to do, I feel sure our religious lives and concerns can only deepen and grow. We may be changed, but I don't see why the changes cannot be changes for the better.

REPORTER: The religious response—in Washington, in Peking, in Cairo—is only just beginning. We have not reported on all of it, but only on that part close to home. It is clear that we are at the beginning of a journey, a journey for which there are as yet neither maps nor clear destinations. Religiously, as we have seen, the journey may well take us beyond where any of us now are: beyond our own religion and irreligion, beyond Jew and Christian, church and scripture, wisdom of the East, folly of the West. Beyond . . . perhaps—to what men have called mysticism.

We have tried to show tonight that the radio signals we have received, which may well alter radically and decisively the lives of all, must be received by a morally sensitive and responsible America. We must deserve the opportunity, for if we do not, the opportunity may turn into a disaster. Religion as institution and creed, as the Rector remarked in his sermon, may turn out to matter very little. But religion as wonder, religion as moral value—these matter very much indeed. By our response and responsiveness we can determine whether we grow or die. Dark Age or Renaissance? It is in our hands.

Thank you for joining us. Good night.

"World views make a difference." This is the central thesis of Ron Huntington's attempt to look through the eyes of a Hindu or Buddhist at the possible existence of ETIs.

"I have the fervent hope that, if extraterrestrials come to this earth, they will have the good sense to land where there is a Hindu/Buddhist culture rather than in our Western culture. Our Western approach has been to make other people come to us and explain themselves, to make them communicate in our language. Invariably, the Eastern approach is to go the extra mile to reach the other person. The stranger is the guest."

Ronald Huntington is an interdisciplinary scholar, a Renaissance man, who currently teaches courses in six different departments: English, Music, Religion, Philosophy, History, and Art. Having lived at various times in Southeast Asia, he is also a "bridger of chasms" between East and West. He is the author of *Understanding the World's Religions* (forthcoming).

Dr. Huntington is Professor of Philosophy and Religion at Chapman College in California.

"Indian thought invites us to think in terms of Space-ship Universe, with its unnumbered inhabitants, human and otherwise. . . . The Isha Upanishad's exhortation still sounds its ringing challenge: 'All this is for habitation. Having renounced, enjoy!'"

Ronald Huntington

Mything the Point: ETIs in a Hindu/Buddhist Context

Then—in that timeless, dimensionless point before the creative act occurred—there was neither existence nor non-existence. All that is was there in potential; no-thing was to be found in actuality. There was neither up nor down, nor earth, nor star-crowded heaven. There was no death, and hence no immortality. This All was as a boundless abyss of water, holding in its undifferentiated and indifferent voidness all the droplets of discrete universes, stars, planets, mountains, rivers, and life-forms.

Was there a movement? Where? Was this infinitely diverse Oneness energized? By what? Truly even the gods came later than the moment of creation. Who, then, can tell out of what the worlds have issued forth? Surely some nameless essence that observes impassively, that breathes quiescent in the deepest recesses of existence knows whence came all these names and forms that enchant us. He alone knows indeed. But then, perhaps he knows not![1]

Thus did an ancient Indian sage speculate some thirty centuries ago about the origins of what we too easily call the existent. And, like all the efforts of man to penetrate the ineffable mystery of how the One becomes the Two and then the Ten Thousand Things, the seer's diligent probing ends indecisively. For the part cannot fathom the whole, either by cognition or intuition. The world of actuality is a world of distinctions, a multiverse rather than a universe. To exist is to distinguish—between I and not-I, earth and heavens, poet and philosopher.

Faced with the imponderable, the Indian mind neither shied away in despair nor settled for human-sized solutions. Maintaining by the assurance of faith the conviction of a oneness underlying and infusing all the seemingly capricious manifestations around us, it refused to limit that unity by earthbound hypotheses or humanly comprehensible boundaries. And as later seers were to work out the implications of the early Vedic vision, we encounter temporal and spatial dimensions that are staggering even to those accustomed to dealing in geologic ages and light-years.

As if with foreknowledge of the inevitable failure of all methodologies founded in the diversity of existence when they are applied to a distinctionless and all-encompassing ultimate reality, these seers expressed themselves in myth, that poetic protoscience that transcends the modern bugbears of consistency and objectivity. That is to say, their awareness of man's inability to consider the entire universe—including *himself*—as an object caused them to speak analogically and psychologically rather than logically. Chaos and order, good and evil, subjectivity and objectivity, time and eternity, and you and I are all relative categories—relative to each other, to be sure, but also relative to the patterns of human thought. The so-called pathetic fallacy, attributing human qualities to nature, and common anthropomorphisms in conceptualizing deities are both unavoidable in the final analysis. For it is man whose mind is involved in the conceptualization, and we are all alike in being victims of what Ralph Barton Perry called "the egocentric predicament," or what the early Indian Jaina logicians and more recent Western physicists have emphasized as "the standpoint of the observer." To ignore this is to engage in self-delusion; to be conscious of it is simply to recognize the poetic, the mythic quality of *all* "explanations."

I

Let us first consider the Indian view of history. Unlike the more familiar linear assumptions that underlie most Western thought, traditional Indian speculation pictures the historical process as an unbroken circle. On the analogy of such natural processes as the succession of seasons, or of night and day following each other, the Indian saw history as beginningless and endless, an uninterrupted chain of cause and effect relationships in perpetuity. And because this circular process has neither germination nor termination in any ultimate sense, any and every moment could be *defined* as a beginning and, coequivalently, as an ending.

An inescapable corollary of this view is that nothing is ultimately created or destroyed, although it may proceed through a dazzling array of transformations. Because of their utility in daily life or their relative longevity, as compared to the brief spans of individual or corporate human existence, many of these transformations—perhaps we could call them incarnations of the fundamental world-stuff—delude us into mistaking the transient form for the permanent substance, the incidental for the transcendental. A time-honored Indian analogy speaks of it as attributing greater reality to the pot than to the clay from which it is made. But modern physics provides still more telling analogies. The law of the conservation of matter has required modification as physicists have demonstrated that matter can be transformed into energy and vice versa. Both matter and energy are now seen as human constructs to deal with an unknown and, according to Werner Heisenberg's principle of indeterminacy, finally unknowable X, which we perceive under first one and then the other label. And that X cannot be ultimately destroyed. Space and time are similarly seen as merely human modes of perception and classification rather than as eternally exclusive categories of nature itself.

We enter inevitably into a world of paradox, a world in which the wave and particle theories of light stand in logical contradiction, yet each possesses immense utilitarian value. The static/dynamic bipolarity of that world has been well captured by the philosopher Alfred North Whitehead: "In the inescapable flux, there is something that abides; in the overwhelming permanence, there is an element that escapes into flux."[2] T. S. Eliot's poetic phrase expresses the same paradox: "At the still point of the turning world, there the dance is."[3] From this perspective, finally, Ecclesiastes' melancholy "Verily, there is nothing new under the sun" is seen as referring with equal validity to the world that the Buddha described as being in a process of perpetual change.

Change and changelessness are the two sides of a coin, double aspects of the human mind's capacity for grasping and "ordering" the universe in which its consciousness exists. Each can be the basis for a self-consistent view of our world, though each denies the other. Yet both, taken *simultaneously*, are true of a world that is neither as orderly as some would insist nor as chaotic as others would portray. Overflowing with meaning, it is at the same time as meaningless as Sartre has pictured it, for these and all other descriptions have only a relative and utilitarian value, to be picked up and then discarded as one strives for greater inclusiveness. Both the ups and downs are part of the self-same circle, and a concentration on either is a denial of the totality. To see birth and growth without in the same experiences recognizing the equal possibility of their being described as death and decay is self-delusion. All are human definitions, created by the power of selective thinking and exercising the bewitching magic that blinds us to their opposites. That which is awe-full is simultaneously awful.

II

Against this all-encompassing vision that ultimately annihilates and incorporates all distinctions, it is still possible for man to make relative judgments. Within his individual set of intellectual coordinates, or even on broader assumptions involving humanity or life in general, it is possible to determine what shall be called values and disvalues. And so, within the immeasurably large circle of process, of being and becoming, the Indian claimed that there exist at any given time an unnumbered variety of superior and inferior "worlds" alongside our earthly habitat. We might call these heavens and hells, assigning each "world" to one or the other of these categories according to our perception of its relative desirability compared to our present existence as earthlings in human form. In other words, "hells" are those forms of existence that are inferior to our own (by our limited lights), and "heavens" are those places or planes of existence that appear to us as superior or more highly evolved. From another angle, just as my subjective "world" at any moment may seem to me more or less desirable than my "world" at another moment, or more or less desirable than that of some other person, so may I project the existence of other worlds in the so-called objective universe.

In keeping with the fundamental law of transformation, which precludes any ultimate creation or destruction of the unknown and unknowable stuff of the universe, the present existence of "me" in this time, place, and form has been preceded and will be followed by continued existence elsewhere in the universal process. Such is the essence of the often misunderstood principle of reincarnation, or transmigration, which is so central to all three of the major religions that grew up on Indian soil—Hinduism, Jainism, and Buddhism.

In short, the Indian philosophical and cosmological framework as found in these religions explicitly assumes the existence of extraterrestrial intelligences, in the most literal meaning of the term, in habitats and modes of being both superior and inferior to human life as we know it. Moreover, that framework provides no inherent reason why you or I should not have been an ETI in a prior existence or could not be one in some future existence. As Huston Smith has said, "If we were to take Hinduism as a whole—its vast literature, its opulent art, its elaborate rituals, its sprawling folkways—if we were to take this enormous outlook in its entirety and epitomize it in a single, central affirmation we would find it saying to man: You can have what you want."[4]

The process by which one's reincarnated status is determined is a totally impersonal one, a law of cause and effect applied not only to the physical world but also to mental and moral processes—the law of karma. And all beings in this universe, whatever their status or form in the space-time continuum—humans or other animals on this planet, "gods," "demons," or under any other rubric—are subject to this law and therefore to reincarnation. The lines between humanity, divinity, diabolic existence, and the animal kingdom are man-made conveniences and are ignored and transcended by the surging tide of universal transformation. Indian mythology is full of tales of humans who, by a lengthy and arduous accumulation of good karma extending over perhaps several lifetimes, achieve rebirth in some higher sphere—as "divinities" by human standards. The reverse also holds true, and there are "gods" who, perhaps through excessive pride of position or power, lose their exalted status, to be reborn in some form in one of the inferior realms. To one of these myths we now turn as an illustration.

III

The story is told in the Hindu Brahmavaivarta Purāna of the god Indra, whose great strength enabled him to triumph over chaotic forces that threatened to engulf all life in the heavens, on earth, and in the nether regions. So flattered was he by the praise bestowed upon him as a result of his mighty victory that his vanity and ambition became boundless. At length some relief was essential from this insatiable egotist.

One day there appeared before him a handsome boy, full of the inno-

cence of youth. Indra ostentatiously heaped gifts upon his guest and proceeded to boast of his prodigious accomplishments. The young visitor listened respectfully, and then said: "Surely, none of the preceding Indras
has been as powerful and successful as you."

The casual reference to former Indras was not lost on the conceited
host, and he indulgently inquired further of this seemingly naïve youth.
"Many indeed are the Indras I have seen," came the response, "one succeeding another, eon after eon, as the great wheel of existence grinds away
inexorably. Worlds come into being and dissolve, much as drops of water
are thrust up by the ocean's waves and sparkle momentarily in the sun-
light, only to return to that formless mass from which they arose. That
which is born in time and space must also die in time and space. In human
terms its lifetime may be no longer than the brief moment of a random
thought. Or it may, like Brahmā, be 1,088,640 times as long as the
seventy-one-eon lifetime of an Indra—but *it will die*, even as one Brahmā
follows another ceaselessly. Who can count the successive worlds that have
arisen and perished? Who can number the Indras who have ruled in a beginningless series, and shall rule afterwards in the endless future?" The
haughty conqueror cringed and his mouth opened as if to speak, but the
mysterious boy continued.

"Hold! I have spoken only of those worlds within this universe. But
consider the myriads of universes that coexist side by side, each with its
Indra and Brahmā, and each with its evolving and dissolving worlds.
Stretching beyond the limits even of your mind, O great Indra, are those
universes. Can you presume to know them, count them, or fathom the
reaches of all those universes with their multitude of worlds, each with its
legions of transmigrating inhabitants?"

Indra's once indomitable pride crumbled before the sheer magnitude
of the vision presented by his young visitor, who, of course, was none other
than an incarnation of that transcendent wisdom hidden deep within each
atom of the universe—the nameless essence spoken of in the hymn with
which this essay began. But that is not the conclusion of the story, for
otherwise Indra and the reader would be left with a total devaluation of all
activity within the time-space arena. The depression that results from
staring directly into the abyss of meaninglessness is not a more accurate
portrayal of reality than the psychic inflation that had preceded it.
Through the mediation of yet another incarnation, this time of a deity
whose function it is to kindle within us a fascination with the diversity
around us, Indra is led to a middle path between the equally undesirable

extremes of over- and under-valuation of existence.[5]

"Man," in Abraham Joshua Heschel's trenchant phrase, "is a creature of magnificent splendor and pompous aridity." It is only by grasping firmly and simultaneously both ends of the paradox that one can experience the fundamental law of existence enunciated in the Isha Upanishad: "All this [the universe] is for habitation. *Having renounced, enjoy!*"[6]

IV

Examples like the foregoing could be multiplied almost without end from the rich mythic tradition of the Hindu and the Buddhist. But the meaning is already clear. Man is one part of a continuum of life that extends not just to the various forms on this planet but to other worlds within this universe as well. And the Purāṇic author of our account goes still further, for he postulates the existence of parallel universes beyond the outermost reaches of our own. As startling and as inherently interesting as that concept is, its implications lie beyond the limits of this essay, and we turn now to the Hindu concept of the avatāra, which is of more direct significance.

Commonly translated as "incarnation," the term *avatāra* literally means "one who crosses over," with the added connotation (from the prefix) of "away" or "down." In other words, the avatāra is one who, having come from some other plane of existence, crosses over or descends into the terrestrial environment. This descent, in the myths of the avatāras, produces a corresponding ascent in the evolution of earthly consciousness.

It is not known exactly when the concept originated in Indian thought. Documents dating from the first few centuries of the Christian era list seven, nineteen, twenty-two, or even more avatāras.[7] It was only through a lengthy and gradual process that the number came to be fixed, by the early eighth century A.D., at ten—nine past and one future. But several of the stories associated with this normative group are found or alluded to in the very early literature, i.e., at the dawn of the first millennium B.C.[8] The manner and brevity of these allusions leads unmistakably to the conclusion that such tales were in general circulation, and we may safely assume that the fundamental concept of the avatāra, though not its later systematization, was an accepted part of Indian thought-patterns virtually at the outset of recorded history.

Several features of the avatāra concept are relevant to our consideration of the possibilities of meeting and accepting visitors from other worlds. First, the one who "crosses down over" is not necessarily to be

expected in human form. In the "orthodox" list of ten avatāras there are included a fishlike creature with a single horn on its nose (Matsya), a half-human and half-lion creature (Narasiṁha), and a dwarf who, however, is able in three strides to take the measure of the entire earth and the heavens (Vamana). Even the apparently human avatāras, for example, Krishna, have attributes distinguishing them from what might normally be regarded as human characteristics. Krishna's skin is customarily portrayed as dark blue, and in his cosmic form as an incarnation of Vishnu he is pictured with four, or sometimes more, arms. These nonhuman forms are paralleled by many other divinities portrayed in Indian art, for example, the pot-bellied elephant with but one tusk (Ganesha), the monkey-god (Hanuman), and the demonic Rāvana, who is pictured with twenty arms. Hindu mythology and art do not lead the Indian to expect visitors from other worlds to appear in any guise immediately accessible to human imagination.

Second, the avatāra's role is not formulated in terms of human ethical systems, nor are his actions inevitably limited by the natural laws known to humans. For example, Paraśurāma's celebrated feat is the destruction "thrice seven times" of the warrior-ruler caste, after which he renounced all forms of violence and departed from the earth.[9] And both the amorous exploits of Krishna and his feats of incredible strength (for example, the lifting of Mount Govardhana) lie outside the canons of customary human morality and physical laws respectively.

Third and most important, the avatāra is eagerly anticipated by the Indian rather than feared. As Sri Aurobindo puts it, "The Avatar is one who comes to open the Way for humanity to a higher consciousness."[10] Were an ETI to "cross down over" into the midst of an Indian festival at the confluence of the Jumna and the holy waters of the Ganges, there is nothing in traditional Indian thought to suggest that the destructive weaponry resulting from work at the great nuclear laboratory at Trombay would be called into readiness. On the contrary, the expectation of Kalkī, the avatāra who is still to come, is such that, while the time of his advent and his external appearance are never made explicit, he is awaited almost in the spirit of the Christian book of Revelation: "Even so, Lord, come quickly." And, we may add, *in whatever form!*

Lest the mythic scheme of avatāras appear to be in conflict with aspects of Indian thought discussed earlier, it should be added that the group of ten applies to the current era, and not to preceding or future ones, in the circular-time theory. Kalkī will arrive to conclude the present age in this universe, not in time itself or other, parallel universes. Other worlds and other eras follow their own processes of evolution and involution, with their own avatāras. This leads to a final observation that it would be an altogether natural extension of the concept to assume that visits by terrestrials to other civilizations could also be satisfactorily explained by the idea of the avatāra.

V

Although the point is fast approaching becoming a platitude, it is nonetheless worth emphasizing that we still tend from long-hallowed habit to overlook the fact that our answers most often inhere in and are predetermined by the manner in which we ask our questions. Those questions themselves presuppose a complex framework—in Jain terms, a *naya*—of presuppositions that occupy the privileged status of "truths we hold to be self-evident." In creating our models of the universe it is easy but deceptive to project them into that universe and then claim them as objective discoveries about an objective reality.

Several centuries ago we were forced (not without a good deal of reluctance) to accept the conclusion that the earth is round, although the curvature of its surface is imperceptible even to the person sitting on a waterfront quay and viewing the horizon. In the twentieth century we are being forced (again with great reluctance) to accept the concept of the curvature of space—that we live in a finite but unbounded universe. The presuppositions that underlie the Indian traditional thought of Hinduism, Jainism, and Buddhism challenge us to consider as well the possibility of the curvature of history and of the human mind itself.

To conceive of the historical process as a closed circle is to relegate to the realm of nonsense many of the questions that have occupied and vexed the greatest minds of Europe and its transatlantic offspring. How (or when) did it all start? What was the "first cause"? Where is all this leading? How will the world end? What, finally, is the meaning of history? These and similar questions, asked by theologian, philosopher, and scientist alike, become meaningless gibberish if the linear model of history is replaced by a circular model. Optimist and pessimist stand equally condemned and pitiable for having allowed themselves only the openness to perceive one half of reality. Yet it appears that the given conditions of human existence deny us the ability to conduct the crucial experiment to determine which, if either, model for history is correct.

Like all else that exists, however, the Indian view of history cannot be adopted without cost. Its price tag may finally be judged exorbitant, at least by the European mind, for it involves no less than the renunciation of any ultimate purpose in the universe, and very possibly of the reality of history itself in exactly the sense that we most wish it to be real. For if time is eternal—and that is a paradoxical yet ineluctable corollary of the circular model for the historical process—then there can be no redemption, human or divine, of either time or eternity. We may thus prefer to continue the age-old and frustrated quest for a surd evil as the counterpart to a deity we have defined as absolutely good rather than to give up the profoundly satisfying conviction that our actions have some ultimate meaning and purpose. It may be that we shall prefer to contend with our moral anxiety, which can at least spur us to increased activity on behalf of some hope for the future, rather than to open the door to ontological anxiety (the terms are Paul Tillich's).

Still, the question may fairly be raised whether the linear view of history to which we are accustomed does not itself come at an exorbitant price. For while it supplies a strong sense of purpose and meaning to existence—and especially to human existence—it does so at the incalculable cost of bifurcating our universe. This is most obvious in the sphere of values and morality, but it cuts through every aspect of philosophy, theology, and, to a large extent, the natural and behavioral sciences. It may be that in attributing an absoluteness to a host of relative distinctions—good and evil, we and they, organic and inorganic matter, humanity and divinity—we have given them a burden that is too heavy for them to bear.

To be sure, the circular view of history is not supportive of the human ego. It supports neither individuality nor total separateness as fundaments of reality. But whether in terms of neighborhood relations, political relations, international relations, or interplanetary relations, the individual and collective human ego may be the greatest threat to openness and acceptance and therefore, in the final analysis, to its own continued existence as the separate entity it struggles to be.

Much the same is to be said concerning the curvature of the human mind. To be conscious of a thought—"I have an idea"—is to make the bifurcation into a subject-object relationship. Thoughts do not exist apart from this easily ignored relationship, and they are inevitably slanted to some extent by ego involvement. The self-preservative and self-aggrandizing nature of the thinking mind is, paradoxically enough, expressed in its predilection for minimizing its own creative role in *designing* the world it pretends only to be describing objectively. In a subtle but nevertheless comparable manner, we are like the freshman psychology student who, on hearing the professor remark that most people take things too personally, cheerfully replied, "I don't!" Ideas come to be invested with a rightness and tightness born of their unconscious coupling with personal value-judgments, and it is tempting to conclude that the "sin of pride" is to be located in man's usurpation of the very function that he consciously attributes to his god—that of creating the world—and all this without sufficient self-awareness to be cognizant of his role as usurper.

The world confronts us in all its logic-annihilating and security-shattering ambivalence, and it seems a necessity to shield ourselves from one half of life in order to maintain some sense of orderliness and inner consistency, too often defined as identical with "sanity." Small wonder that the unknown or "irrational" is feared, that ETIs pose a feeling of discomfort, and that even the possibility of extraterrestrial visitation has elicited statements and reactions that would only be described as paranoid were it not that they harmonize with the particular paranoia of the cultural thought-patterns that surround them.

The message of the myth of Indra is that the great chain of life is unbroken, both in and beyond the abode of earthlings. The message of the avatāras is that the nameless essence underlying all manifested forms is capable of infinite diversity and that there is nothing to fear from ETIs or any of the multitudinous forms of that essence.

Psychologists and sociologists alike have called attention to the essential loneliness of our inner world as we spin through the momentary span of our individual lives. In many of our novels and poems, as well as in the contemporary theater, we encounter extended parables on this theme. Cartoonists like Abner Dean have caricatured the idea, often to our vague discomfort. Buckminster Fuller has significantly broadened the image to encompass all the human passengers aboard Spaceship Earth. The expansive message of traditional Indian thought invites us to extend these as yet microcosmic analogies to a macrocosm almost beyond our conception and begin to think in terms of Spaceship Universe, with its unnumbered inhabitants, human and otherwise. And in the immensity of that universe —self-enclosed, yet only one among an unknown and forever unknowable many—the Isha Upanishad's exhortation still sounds its ringing challenge: "*All* this is for habitation. Having renounced, enjoy!"

NOTES

1. A free rendition of the Nasadiya Sukta (Rig Veda X. 129).

2. Alfred North Whitehead, *Process and Reality* (New York: Macmillan, 1929), p. 513.

3. T. S. Eliot, "Four Quartets," in *Collected Poems, 1909-1962* (New York: Harcourt, Brace and World, 1963), p. 177.

4. Huston Smith, *The Religions of Man* (New York: New American Library, 1958), p. 25.

5. *Brahmavaivarta Purāṇa* (Calcutta: Gurumandal Series No. 14, 1954-55), II, 836-43. The myth is related in greater detail in Heinrich Zimmer, *Myths and Symbols in Indian Art and Civilization* (New York: Harper & Row, 1946), pp. 3-11.

6. *Isha Upanishad*, 1.

7. For these lists see Vāyu Purāna 97:137-42, Garuda Purāna I.202, Bhāgavata Purāna I.3:26-27, etc.

8. For example, the Matsya avatāra may be prefigured in Rig Veda VII.88.3-5, and appears in developed form in Śatapatha Brāhmaṇa I.8.1. The Vāmana avatāra is clearly alluded to in Rig Veda I.154.2 and Śatapatha Brāhmana V.2.5.4.

9. Bhāgavata Purāṇa IX.15:16-16:27.

10. Sri Aurobindo, *On Yoga, Book Two*, I (Pondicherry: Sri Aurobindo Ashram, 1958), p. 413.

Would ETIs necessarily be persons? It's a good question, and if it is true that "persons" are the proper objects of ethical concern, then if encounter occurs the question will become critical.

Michael Tooley has previously attempted to establish rational criteria for judging whether or not living things are also persons. Needless to say, the question has immediate practical implications for a number of current ethical problems—abortion, euthanasia, the right-to-die cases in our courts, and related ethical problems. Tooley made such an inquiry—with significant results—in an article entitled "Abortion and Infanticide," *Philosophy and Public Affairs*, vol. 2, no. 1 (1972).

Dr. Tooley was Professor of Philosophy at Stanford University before he became a full-time research scholar at The Research School of Social Sciences at the Australian National University.

"Contact with ETIs would raise the ethical problem of how we ought *to interact with them, and the ethical problem is a matter of providing an answer to the question of what makes something a person."*

Michael Tooley

Would ETIs Be *Persons?*

Contact with a society of intelligent extraterrestrial beings would compel us to grapple with some exciting and disturbing questions. Two areas would be especially important. First, such an alien society would force us to reconsider some of our most basic beliefs, values, and attitudes and to inquire whether those fundamental beliefs and values were really justified. Second, it would raise the ethical problem of how we *ought* to interact with extraterrestrial, nonhuman intelligences.

CULTURE SHOCK AND EXTRATERRESTRIAL BEINGS

What are some of the ways in which confrontation with societies whose basic beliefs and values are radically different, not only from those of our own society but from those of any human society, might force us to re-examine our fundamental beliefs and values? Among the areas that would seem most exposed to such culture shock are religion, our attitude toward death, our patterns of human interaction, and our attitudes toward the improvement of man.

An extraterrestrial society of intelligent beings with a highly developed culture and an advanced scientific outlook, but with no religious orientation, would make it difficult to avoid questioning both the supposed human need for a system of religious beliefs and the rationality of a religious outlook.

A society where the problem of aging does not exist, either because of the different biological nature of the organisms or because a solution to the problem of aging has been discovered, would compel us to call into question the almost universal acceptance of death as a fact of life in human societies.

Confrontation with societies whose social organization is radically different from ours, but in which people seem to be functioning effectively and happily, would force us to ask just how satisfactory our present institutions of marriage and the family and education really are.

Finally, a society that seems admirable to us and in which improvement in the nature of its members has been achieved through the application of scientific knowledge to the production and raising of offspring would call into question our fear of such an approach to the problem of improving the nature of man.

All of these questions can, of course, be raised at any time. But in the absence of societies that differ radically from our own, we often evade these questions, on the grounds that the alternatives suggested either are not really possible or, if possible, would lead to disasters of various sorts. Exposure to actual societies where these alternatives are lived and where there are no associated undesirable consequences would render such evasion impossible.

ETHICS AND INTERACTION WITH ETI'S

Let us turn now to the second area—the one on which I shall focus in this essay—the problem of how we ought to interact with a society of extraterrestrial intelligences. The central question here is simply whether it would be morally permissible to treat them in any way we wanted or whether, on the contrary, certain ways of treating them would be morally wrong. In our interaction with things in everyday life we generally divide the world into three classes, and the ways in which it is morally all right to treat something depend upon which of these classes it falls into.

First, there are inanimate objects, plants, and very low forms of animal life. The common view is that absolutely no behavior is wrong in itself with respect to things belonging to this class. It may be wasteful to destroy inanimate objects, especially complex machines, but we do not do wrong *to* those objects in destroying them. If such destruction is wrong, it is because someone is deprived of their use. Similarly, if a person thinks it is wrong to destroy a forest, the reason generally offered is that in so doing one is depriving other humans, or higher animals, of their use, not that one is harming certain plants.

Second, there are animals that, although subhuman, seem sufficiently developed to be capable of experiencing suffering. The common view here is that, while there are *some* ways of treating such organisms that involve doing wrong *to* them, some behavior that would be seriously wrong in the case of humans is not so in the case of these nonhuman animals. Thus it is generally thought to be seriously wrong to torture a cow for the pleasure of doing so, but not seriously wrong to kill one for the pleasure of eating it.

Finally, there is the class of normal adult human beings. It is here that there is the greatest restriction upon the scope of treatment that is ethically acceptable. The critical contrast, for our purposes, is that, while members of the first class may be destroyed for no reason at all and members of the second class may be destroyed for quite insignificant reasons, only the most serious grounds can make it morally permissible to destroy a normal adult human being.

These three classes provide three possible models of what behavior might be appropriate with respect to extraterrestrial intelligences. Is anything permissible with respect to extraterrestrial intelligences? Or would it be permissible to destroy them for quite trivial reasons but not permissible to do some other things to them? Or would it be (at least) as seriously wrong to destroy an intelligent extraterrestrial being as to destroy a normal adult human being?

Providing a satisfying answer to this problem is quite difficult. We will see that it raises many quite difficult issues in diverse areas of philosophy, specifically, in ethics, in philosophy of mind, and in theory of knowledge. And when we pursue those issues we will see that there is no single answer that is correct for all conceivable intelligent extraterrestrial beings. What behavior is morally acceptable will depend upon certain additional facts about the nature of the extraterrestrial beings in question.

THE RELEVANCE OF THEORY OF KNOWLEDGE
AND PHILOSOPHY OF MIND

How is it that a question in ethics can lead one into questions in other areas of philosophy, such as philosophy of mind and theory of knowledge? Consider the question of why it is wrong to kick dogs but not wrong to kick trees. The natural answer is that dogs are capable of experiencing pain while trees are not. So far there is no problem. But, one might ask, How does one know that dogs can experience pain while trees cannot? This question raises an issue in theory of knowledge—the issue of how one can know what individuals enjoy states of consciousness and what specific states those individuals are capable of being in. And it is important to realize that the worry here is not one that should trouble only philosophers. Some recent experiments dealing with the reactions of plants to emotions suggest that our everyday belief in the unconsciousness of plants is by no means unchallengeable.

Issues in philosophy of mind arise if one goes on to ask what it is to be conscious and what it is to feel pain. Philosophers have given very different accounts of what the mind is and of the nature of particular mental states, such as the experience of pain. Some philosophers, called behaviorists, hold that to say that an organism has a mind is just to say that it is capable of behavior of a type or complexity that sets it off from inanimate objects and from living things, such as plants, that have no minds. Other philosophers, generally referred to as identity theorists, maintain that the mind is, as a matter of fact, identical with the brain, so that mental states are states of the central nervous system. Finally, there are dualists. They maintain that the mind is something nonphysical and private, so that the only mental states one can ever observe are one's own.

These differing opinions about the nature of the mind do not affect our ordinary ethical decisions. Behaviorists, identity theorists, and dualists generally agree about what things are conscious and what things are not. But while these differences are unimportant in the context of everyday life, they do not remain so in a biocosmic perspective. The importance of these disagreements about the nature of the mind can perhaps best be brought out by supposing that at some future time we succeed in making a robot whose behavior is indistinguishable from human behavior. What class would such a robot fall into? Would it be just a complex inanimate object that one could treat however one wanted? Or would it have sensations, thoughts, and feelings, so that it would be wrong to treat it in certain ways? If its behavior were indistinguishable from human behavior, a behaviorist would have to say that such a robot enjoyed all the mental states that humans enjoy. An identity theorist would, I think, be forced to take the same view. A dualist, on the other hand, would probably say that the fact that the robot's behavior was exactly like human behavior did not provide sufficient reason for holding that it enjoyed states of consciousness, conceived of as states that are nonphysical and private. The upshot is that the dualist would probably think it permissible to do anything to such a robot, whereas the behaviorist and the identity theorist would hold that anything that it would be wrong to do to a human would also be wrong in the case of the robot.

One consequence deserves to be emphasized, namely, that what appears to be general agreement about central ethical principles is in some cases only apparent agreement. For while virtually everyone will say that pain is bad in itself, what one is really asserting will depend upon one's view of what it is to be in pain. For the behaviorist pain is a matter of how one is behaving or disposed to behave. For the identity theorist it is a certain process in the brain. And for the dualist pain is something nonphysical. So actually, people mean very different things when they say that pain is bad, and the agreement is only apparent agreement. Appreciation of the underlying disagreement is critical in the present context, since it has a fundamental bearing upon the answer one will give to the question of how intelligent extraterrestrial beings ought to be treated.

THE ETHICAL ISSUES

Let us consider the fundamental ethical issues that must be settled in order to determine what behavior would be morally acceptable with respect to extraterrestrial beings. The basic questions to be answered are these: (1) What sort of beings is it possible to injure? (2) What sort of beings is it seriously wrong to destroy?

The first question can be dealt with quite briefly. To begin with, one surely wants to say that only beings that are capable of consciousness can be injured. Does one also want to say that this is sufficient, so that any being that enjoys consciousness can be injured? I think not. Imagine an extraterrestrial organism that is a pure perceiver, that is, an organism that has sensory experiences of its environment but has absolutely no interest in anything and is incapable of experiencing pain. Leaving aside for the moment the possibility of destruction, is one not inclined to say that there is nothing that one could do to such a being that would injure it in any way? If so, then sheer consciousness cannot be enough. It is possible to injure something only if it is capable of experiencing pain, or at least of having desires that can be frustrated.

I think that this answer would be accepted by most people. But again it needs to be emphasized that the agreement here is in large measure apparent. In the first place, the behaviorist, the identity theorist, and the dualist will usually assign quite different interpretations to the notion of having desires. And second, there may be important disagreements even among those who share the same general outlook on the nature of the mind. Some dualists, for example, would be content to view desires simply as states that serve to explain behavior in a certain way, while other dualists would insist that desires, even if not themselves states of consciousness, must stand in some relation to states of consciousness. So even dualists may well have serious disagreements about what sort of beings it is possible to injure.

THE CONCEPT OF A PERSON

Let us turn now to the second, and much more difficult, question: What type of beings is it seriously wrong to destroy? One way of approaching this issue is to consider organisms whose destruction one does regard as seriously wrong and to ask why it is seriously wrong to destroy such organisms. So let us ask why it is seriously wrong to destroy normal adult human beings. The answer often given is: simply because they are *humans*. When one thinks only of the living things that one encounters on our own planet, such an answer is tempting, since humans are, as a matter of fact, the only individuals to which most people would assign a serious right to life. One advantage of the biocosmic perspective is that the parochialism of this answer is immediately apparent. It is conceivable, for example, that there are nonhuman animals on Mars that speak languages, have highly developed cultures, have advanced further scientifically and technologically than humans have, and who attribute sensations, thoughts, and feelings to themselves and to us. In such a case it would surely be as wrong to kill Martians as it is to kill humans. So it cannot be membership in a particular biological species, such as Homo sapiens, that makes it wrong to kill something. Instead, the reason appears to be that it enjoys certain psychological states and/or capacities—states and capacities shared by humans and the Martians we are imagining. If, as seems convenient, we employ the term *person* to refer to an entity that possesses the relevant psychological properties, we can say that the basic moral principle involved here is that it is seriously wrong to destroy persons.

But what exactly is it that makes something a person? This is the question with which we must now grapple. It is surely true that something cannot be a person unless it enjoys mental states, in particular, states of consciousness. But this is not sufficient. So we must ask what *more* is required beyond the capacity for consciousness to make something a person.

One of the fundamental issues to be resolved in answering this question is whether the something more is a matter of an entity's relations to or interactions with other individuals, or whether it is instead simply a matter of certain properties that the entity has independent of its relations to others. Some have suggested that to be a person is to be a conscious being that enters into social relationships with other conscious beings or, at least, that has the capacity to enter into such relationships. In contrast, others have maintained that what makes a conscious being a person is the capacity for rational thought—something that it would seem that one could possess independent of any relations with other conscious beings.

Is something like the capacity to interact with others essential to our concept of a person? It might seem at first that it is. But what of a normal adult human being who suffers complete paralysis? If we have reason to believe that the brain damage has not impaired his ability to think and feel, we will surely want to hold that it would be wrong to kill him, even though he no longer has any capacity for social interaction. In the case of the persons with which we are familiar such cases are exceptional. However, there might well be species of extraterrestrial beings who, though highly intelligent and capable of experiences and feelings, are unable to communicate and interact with others. If such beings enjoy a mental life comparable with or superior to our own, it would surely be seriously wrong to destroy them.

It looks, then, as if being a person is a matter of something "internal" —a certain sort of mental life—rather than a matter of an individual's relations to other individuals. But what sort of mental life? What type of mental activity, in addition to sheer consciousness, is necessary if something is to be a person?

One popular suggestion is that it is rationality, or the capacity for thinking, that transforms a merely conscious being into a person. This is a commonly held view and one that may seem initially plausible. However, its initial plausibility rests to some extent upon the vagueness of the notion of thinking. What counts as thinking? Does one have thinking wherever one has "insightful learning," as contrasted to mere trial-and-error learning? If so, chimpanzees are certainly capable of thinking, and thus we must say, on the present suggestion that it is thinking that makes something a person, that chimpanzees are persons and that it would accordingly be seriously wrong to kill them.

There is, however, a much more serious problem. Suppose that we encounter an extraterrestrial organism that has sense experiences and that feels pleasure and pain and whose intellectual capacities far exceed our own. Given the present suggestion we would certainly have to say that such a being is a person. But suppose that, although it speaks English, among other languages, it never utters sentences containing the word *I* and seems incapable of learning to use such sentences. Would not this fact strongly incline us to say that it was probably not a person, its outstanding problem-solving and other intellectual abilities notwithstanding? Such a case seems to show that the possession of certain cognitive abilities cannot be the key to what it is to be a person.

Why would an organism that was otherwise capable of using language be incapable of learning to use sentences containing words such as *I*? What does such an organism lack? The natural answer, I think, is that it must have no sense of any unity of experiences over time. There may be experiences associated with the organism as a continuing physical entity at different times, but either these experiences are not linked together in the appropriate way or there is no awareness of the interconnection of the experiences.

Let us use the term *self* to refer to a being whose experiences and other mental states are linked together in the way that is necessary if something is to be a person. What is it that transforms a mere collection of states of consciousness into a self? A common view that seems quite plausible is that memory plays a crucial role in this unification of experiences into a self that endures through time. The reason that certain experiences that I had yesterday belong to the same self as certain experiences I am having today is that my present experiences are accompanied by memories that are about the earlier experiences. However, it will not do to say that two experiences belong to the same self only if they are directly linked by memory, since no one remembers all of his past experiences. The most natural way of dealing with this difficulty is to make use of the notion of an *indirect* connection. Then one can say that two experiences belong to the same self if one can find a chain of experiences and memories, each link of which consists of some experiences and some concomitant memory that is about some earlier experience that serves to connect up the two experiences in question.

What is it to remember a previous experience? In one sense any organism that learns from experience, in the sense of modifying its behavior in accordance with past experience, might be said to remember experiences. However, it seems to me that mere modification of behavior in response to experience is not in itself sufficient to provide the unity needed to constitute experiences and other mental states into a continuing self. The unity wanted must itself be somehow represented in consciousness. For this, a stronger sense of remembering is required, namely, memory involving dispositions to have *thoughts* about the past.

The suggestion, then, is this. A self is a set of experiences and other mental states existing at different times and linked together by memory understood as involving dispositions to have thoughts about previous experiences. If, then, an individual cannot be a person unless it is a self in this sense, we can say that it cannot be seriously wrong to destroy something unless the thing is a conscious being possessing some capacity for thinking, together with some conception of events occurring in a temporal order.

It seems plausible that something cannot be a person unless it is a self. But is a self automatically a person, or is something more required? One view is that in order to be a person one must not only be a self, but must possess self-consciousness, where self-consciousness is construed simply as one's recognition that one is a self. Thus it is sometimes said that the reason it is seriously wrong to kill normal adult human beings but not seriously wrong to kill cows and chickens is that the former possess self-consciousness, while the latter possess only consciousness, and not self-consciousness. But why should recognition of the fact that one is a self be so critical? Why is not the fact that one is a self itself sufficient grounds for being ascribed a serious right to life?

ONE POSSIBLE APPROACH TO THE PROBLEM

How does one decide this issue? How does one determine whether it is the property of being a self or the property of being self-conscious or some other property that makes it seriously wrong to destroy something? One might appeal to "moral intuitions," though it is far from clear why agreement of moral intuitions of different people serves to justify moral principles, in view of the fact that most moral intuitions seem to have developed through a process of socialization. But even waiving this general point, the appeal cannot really be helpful in the present case, since the moral intuitions of people just do not exhibit any substantial measure of agreement on the question of what it is that makes it seriously wrong to destroy certain organisms.

I feel that there is virtually no hope of arriving at a satisfactory answer without first investigating the general question of how, if at all, ultimate moral principles are to be justified. In the absence of such an investigation it seems unlikely that one can provide convincing grounds for preferring one answer to another. However, an investigation of the foundations of morality is impossible here. What I want to do instead is to set out one approach that seems quite plausible. This approach will involve the formulation of a more general ethical principle from which the injunction against killing persons, as well as certain other moral principles, can be derived. The derivation of the prohibition against killing from the more general ethical principle will automatically entail an answer to the question: What is it that makes something a person?

So let us broaden our base by considering moral principles other than the principle that one ought not to destroy persons. Another basic principle to which most people are deeply committed is this: *It is wrong to inflict pain upon conscious beings, regardless of whether or not they are persons.* It would be nice if one could find some more general principle or principles from which one could derive both this principle, which is not restricted to persons, and the principle that it is wrong to destroy persons. What might such a more general principle be? One familiar and plausible candidate is the following: *The frustration or nonfulfillment of desires and preferences is always an intrinsically undesirable state of affairs, the undesirability being proportional to the strength of the desire or preference in question.* An individual who accepted this as a basic moral principle could attempt to use it to justify both the claim that it is wrong to inflict pain upon conscious organisms, regardless of whether they are persons, and the claim that it is wrong to destroy persons. In the case of the former the account would turn simply upon the fact that pain is not only a state that a conscious organism desires to be free of, but also a state that hinders an organism in its action to satisfy other desires.

But what about the injunction not to kill persons? How might it be derived from the principle dealing with fulfillment and nonfulfillment of desires? One possible approach would be that the reason that it is wrong to destroy persons is that they are *things that can envisage a future for themselves and that have desires about those future states of themselves*, and thus killing such entities is wrong because it prevents the satisfaction of the individual's desire for continued existence.

This provides an explanation of why it is wrong to kill some animals, such as normal adult human beings, but not wrong to kill others, such as oysters. In the case of the former there is a desire to continue to exist as a conscious subject of experiences and other mental states, and killing is wrong because it prevents this desire from being satisfied. Killing oysters, in contrast, is not wrong, so long as it is done painlessly, since oysters presumably do not possess a concept of themselves as continuing subjects of experiences and so are incapable of having a desire to go on living that will be denied satisfaction if they are killed.

This suggested derivation of the prohibition against killing persons provides, in effect, an account of what it is to be a person in the morally relevant sense: something is a person if and only if it is a continuing subject of experiences and other mental states that can envisage a future for itself and that can have desires about its own future states.

To be capable of envisaging a future for oneself as a subject of experiences and other mental states, one must be capable of having the concept of a continuing subject of experiences and of recognizing oneself as such a continuing subject. But this is just to say that one must at least have the *capacity* for self-consciousness. So the present approach, even though it does not show that one must actually be self-conscious to be a person, does show that it will not do merely to be a self. One must at least be capable of recognizing that one is a self.

But what if one were to encounter extraterrestrial organisms that, although they possess both consciousness and self-consciousness, do not have any desires at all—and hence no desire for continued existence. Would they be persons? Would it be seriously wrong to destroy them? If the approach suggested in this section is correct, the answer is that they would not be persons in the morally relevant sense, and it would not be seriously wrong to destroy them. Self-consciousness is a necessary condition of an individual's being a person, but not a sufficient one. The individual must also be capable of having desires about his own future states. In the absence of such desires his destruction cannot be seriously wrong.

What are we to say of this conclusion? Is it "counterintuitive"? It may seem so to some, but this may just reflect the fact that all the individuals we have encountered on this planet who have possessed self-consciousness have also been capable of having desires about their continued existence, and we have not really reflected in any serious way about the treatment of individuals who possess self-consciousness but have no capacity for having desires about their own continued existence. I would suggest that serious reflection would support the view that the capacity to have a desire for continued existence is of critical importance. One reason for thinking so is the fact that in the case of individuals who are capable of having such a desire, the presence or absence of the desire makes a radical difference: suicide and assisting a person in euthanasia are very different from murder.

THE IRRELEVANCE OF LIFE AND INTELLIGENCE

What properties would an extraterrestrial being have to possess in order to be a person? The answer I have offered here is that it would have to be a conscious being possessing both the capacity for self-consciousness and the capacity for having desires about its own continued existence. This leaves us with the question of the conditions under which we would be justified in believing that an extraterrestrial being possessed these properties. But before proceeding to consider that issue, I should like to emphasize two important negative implications of the answer given.

The first is that intelligence, construed in terms of aptitude for solving problems of various sorts, is not essential to the concept of a person. If one applied the term *intelligent* to any being capable of thinking, then it would be true that only intelligent extraterrestrial beings could be persons. Intelligence in that minimal sense does seem essential to the concept of a person, since in the absence of the ability to think there could be no capacity for recognizing oneself as a continuing subject of experiences, nor any capacity for having desires, where desires are something more than mere behavioral dispositions. But aside from this minimal ability to entertain thoughts, intelligence is irrelevant. An organism that could conceive of a future for itself and have desires about its continued survival would be a person, even if it had absolutely no problem-solving ability at all. Conversely, as we argued earlier, there might be extraterrestrial beings with problem-solving ability far surpassing our own that would not qualify as persons, either because they are not conscious at all or, if conscious, because they lack the capacity for self-consciousness or the ability to have desires about their own continued existence.

The second point is that an extraterrestrial being need not be alive in order to count as a person. In one sense of the term, of course, anything that is conscious and self-conscious and that has desires will certainly be classified as alive. But if we understand the term *life* in a biological sense —that is, in terms of capacities such as the ability to reproduce and to collect energy for survival and self-repair—then there might well be extraterrestrial beings that are persons even though not living organisms. There might, for example, be conditions under which we would attribute consciousness, self-consciousness, and a desire for continued existence to robots that had been manufactured by some extraterrestrial intelligences, even though the robots in question had no capacity to repair or reproduce themselves.

THE DECISION PROBLEM

I have argued that to be a person an extraterrestrial being would have to be a conscious entity with the capacity for self-consciousness plus the capacity for envisaging a future for itself and for having desires about its continued existence. The question that we are now left with is how one could determine that an extraterrestrial being possessed these properties, how one could be sure that it was, in fact, a person.

In dealing with this epistemological question, one's view of the nature of mind looms large. If mental states are, as the behaviorist maintains, nothing more than behavioral states—both actual behavior and behavioral capacities and dispositions—then there is no difficulty in principle about determining whether some extraterrestrial being, possibly very different from anything on earth, enjoys mental states, including those essential to a person. One has merely to observe the individual's behavior.

Similarly, there is no problem if the identity theorist is right in holding that the concept of a mental state is just the concept of a state that stands in certain causal relations to behavior and to behavioral dispositions and capacities.

A serious problem arises only if one accepts the dualist view that mental states are states that are not even in principle publicly observable. So the question that must be pursued is whether, if the dualist is right and mental states are private to the individual enjoying them, there is any way to determine whether some extraterrestrial being is a conscious entity possessing the capacity for self-consciousness and the capacity for having desires about its own continued existence, and thus is a person.

Well, what account can the dualist give of our everyday claim to know that other normal adult human beings are persons? If he can account for this, can he perhaps transfer the account to the case of extraterrestrial beings?

The answer is that there are different accounts that the dualist can offer of our claim to know that other humans are persons. Some of these accounts can be transferred, but at least one cannot. So whether it is possible from a dualist perspective to determine whether extraterrestrial beings are persons depends upon which account of our knowledge of other human minds is correct.

The most famous dualist attempt to account for knowledge of other human minds is the argument from analogy. The basic idea is that one can discover, simply from observing oneself, certain laws relating the physical states of one's body to mental states. Thus I notice, for example, that whenever a piece of sugar is placed on my tongue, I have taste sensations of a certain sort. Then, having noticed that other humans are physiologically similar to me in the respects that enter into the laws I have discovered in my own case relating physiological states to mental states, I make the bold move of extrapolating those laws from my own case to the case of all other humans. I can then use the generalized laws to conclude that when there is a piece of sugar on the tongue of another human (and that human is not asleep, etc.), there is a person associated with that other human body that is enjoying a certain sort of taste sensation.

A number of doubts have been raised about the argument from analogy, the most serious of which concerns the legitimacy of extrapolating laws that have been found to hold true in the case of one human body to all other human bodies. But we can ignore these problems here, since even if the argument is sound it will not provide us with any solution to the problem of determining what extraterrestrial beings are persons, except in cases where such beings are, at bottom, very similar to us. For the laws that one discovers in one's own case are, ultimately, not laws relating behavior to mental states but laws relating internal physiological states—specifically, states of the brain—to mental states. Thus one can only apply those laws to organisms whose central nervous system is similar to ours in the relevant respects.

So this method will provide one with no help in determining whether an organism that has a brain with a different structure or that is composed of different chemicals has the mental states that make something a person, or even any mental states at all.

There are, however, other accounts of our knowledge of other minds that are available to the dualist and that *can* be used with organisms whose physiology differs radically from our own. Two approaches are worth mentioning.

First, suppose that we encounter extraterrestrial beings whose behavior apparently cannot be explained in purely physiological terms but can be explained by postulating unobservable states that stand in the same causal relations to each other and to the organism's observable physiological and behavioral states as our mental states stand in to each other and to our observable physiological and behavioral states. The postulation of such states would surely be justified in view of their explanatory power, and if we could not succeed in identifying the postulated states with any physical states, we would surely be justified in concluding that the extraterrestrial beings in question enjoyed states that were mental ones in the dualist sense of being states not open to public observation.

The second approach focuses upon linguistic considerations. Suppose that we meet extraterrestrial beings that speak languages in which we find some terms that it seems appropriate for us to translate by expressions referring to feelings, sensations, emotions, thoughts, and other mental states. Suppose further than the extraterrestrial beings in question are apparently dualists, that is, they draw a sharp distinction between terms that refer to physical objects and their publicly observable properties and terms that it seems natural to translate as mentalistic expressions; and they maintain that the latter expressions refer to events that are not open to public observation.

It would seem that linguistic behavior involving such terms might give us two grounds for attributing states of consciousness to them. First, they might simply claim that they enjoy states of the sort in question, and if we find that they are honest in areas that we can check out independently, we will have at least some reason for believing that they are also telling the truth when they attribute nonphysical states to themselves.

Second, how are we to account for the presence of such mentalistic terms in their language? The presence of such terms in their language will be very mysterious if they never in fact enjoy any states of consciousness.

At this point, however, it may make a difference whether or not the **extraterrestrial beings are living organisms. I have argued that being alive** is not part of our concept of a person; so even a robot might be a person. However, when the question is not one of what makes something a person but of the conditions under which we would be justified in believing that something was a person, the situation is different. Now it becomes relevant whether or not the being is a living organism. If what we are confronting is a robot, it need not be true that the simplest explanation of the fact that its language contains mentalistic terms is that it enjoys states of consciousness. An alternative explanation that is surely simpler is that the person who constructed the robot chose to program into it a language containing mentalistic terms.

This alternative explanation is not available when we are dealing with living organisms that have acquired language by a process of social evolution. Here the most plausible account is that the reason their language contains mentalistic terms is that those terms are necessary for the description of states that they themselves actually enjoy. So if the extraterrestrial beings are living organisms rather than robots, the presence of mentalistic terms in their language would seem to constitute strong grounds for attributing states of consciousness to them. If the mental states that we are thus justified in attributing to them include a capacity for self-consciousness and a capacity to envisage a future for themselves and to have desires about their own continued existence, then they will have to be classed as persons like ourselves.

What I have attempted to do here is, first, to indicate the issues that one must grapple with if one is to answer the question of under what conditions one would be justified in viewing an extraterrestrial being as a person, and second, to outline what seems to me the most plausible answer to this question. The issues raised are difficult ones, falling within a variety of areas—ethics, philosophy of mind, and theory of knowledge. The ethical problem is a matter of providing an answer to the question of what makes something a person, and then of offering a defense of that answer. The second task is critical, since there are a number of alternative accounts available. But it is also difficult, since it seems to carry one into questions about the foundations of morality. All that I have done here is to show how one might defend one answer to the question of what makes something a person by appealing to a more basic and more general moral principle dealing with the satisfaction and frustration of desires.

Issues in philosophy of mind arise due to the fact that the answer to the question of what a person is will be formulated in terms of fundamental psychological terms, and the interpretation assigned to these terms will depend upon what general account of the nature of the mind one accepts. This is apparently of no moral consequence for our own planet, but it would become critical in the case of extraterrestrial beings: one's view of what counts as a person depends upon whether one is a behaviorist, an identity theorist, or a dualist.

Finally, there is the epistemological problem of under what conditions one would be justified in believing that an extraterrestrial being possessed the psychological properties that make something a person. This is not a serious difficulty for the behaviorist or for the identity theorist, but it is for the dualist, in view of the dualist claim that mental states are necessarily private. I have argued, however, that this problem can be dealt with since there is a plausible account of our knowledge of other minds that the dualist can appeal to, both in the case of humans and in the case of extraterrestrial beings.

In conclusion, then, there appear to be conditions that would justify the dualist, as well as the behaviorist and the identity theorist, in believing that certain extraterrestrial beings possess the capacity for self-consciousness together with the capacities for envisaging a future for themselves and for having desires about their own continued existence. And if the account I have offered of what it is to be a person is essentially correct, this means that such extraterrestrial beings would have to be regarded as persons.

Discoveries in astronomy have gradually displaced man from an earth-centered cosmology to a sun-centered cosmology to the truly open de-centered cosmology of today.

Ron Smith suggests that these earlier shifts in perspective, painful as they were, will be minor in comparison to the upcoming dislocation brought about by an encounter with highly intelligent, extraterrestrial life-forms.

Does that mean that you're fearful about such contact?

"No, I'm eager to see it happen. I'm now convinced that a truly advanced civilization, capable of interstellar communication, will have had to pass through a phase in which their destructive tendencies were eliminated or replaced with a higher-order value system. Contact with such ETIs could greatly accelerate our intellectual and, more importantly, our moral growth. We humans could only gain from such an encounter."

Ron Smith teaches astronomy and is Director of the Planetarium at Santa Monica City College. He is the author of numerous articles on lunar and planetary photography.

"After centuries of systematically flushing man from the center of the universe, astronomers have not yet undermined his intrinsic importance. But recently astronomers have undertaken what might be called the 'royal flush.'"

Ronald Smith

The Abdication
of Human Intelligence

For ancient man the earth encompassed the entire universe. The sun, moon, planets, and stars were objects, not places. All of these "objects" appeared to move around a stationary, flat earth. They were continually assembled as they rose above the eastern horizon. They were seemingly disassembled as they set below the western horizon.

Early Greek philosophers considered the universe an extension of natural earth elements. The world consisted of earth and water, while the heavens consisted of air and fire. The earth seemed to exist in a state of constant change, while the heavens seemed totally resistant to change. Thus, of the four basic elements—earth, water, air, and fire—earth and water were changeable, or corruptible. Air and fire were unchangeable and incorruptible. Considering the entire universe, the heavy elements, earth and water, sank to the center, forming the world, which existed in a corruptible state. The light elements, air and fire, surrounded the world, forming the heavens, which existed in an eternal, incorruptible state.

Remarkably, many Christian churches have adopted this Greek cosmology. The Bible never claims that hell is a hot, undesirable place of exile located below the earth's surface. Neither does it claim that heaven lies somewhere on the tops of ethereal clouds. The notion that heaven is up and hell is down derives partly from the Greek concept of incorruptible, eternal elements above, and corruptible, changeable elements below.

Greek philosophers continued to refine their concepts of the earth-centered universe. They observed that the sun, moon, and the five planets that can be seen with the naked eye moved with respect to all the other heavenly objects. No man-made model could successfully account for these observed motions. Thus, out of all the thousands of heavenly bodies, the sun, moon, and planets took on special importance.

With their behavior unexplained, the sun, moon, and planets seemed to possess superhuman attributes. Being constructed of incorruptible air and fire, the planets were viewed as gods. All of these gods moved capriciously around a fixed, centrally placed earth. Obviously then, the planets had to be gods specially ordained to govern significant events on earth. With the planets ruling as gods over earth, their rule, logically, had to influence human behavior. This belief in planetary influence over human behavior initiated the study and practice of astrology.

Throughout the history of astrological practice astrologers have tried to answer two fundamental questions. At what time in a person's life are planetary influences greatest? What are the precise influences of each individual planet? Different answers to these questions have started different "schools" of astrology. Yet all of these astrological "schools" have originated from three commonly held assumptions.

1. The universe is earth-centered.
2. The planets are gods, not places.
3. The planet-gods influence human behavior.

Currently hundreds of millions of people turn to astrology as today's scientific religion. They perceive the practice of astrology as part of "today's" generation. How ironic that "today's" generation chooses to place its faith in scientifically incorrect assumptions over 2,500 years old!

How has mankind come to reject the concept of an egocentric or earth-centered universe? A series of astronomical observations and theories over the past four centuries has finally removed the earth from any special cosmological placement:

1. *De Revolutionibus* (1543). This famous work by Copernicus presented a sun-centered model of the universe to explain planetary motion. Significantly, Copernicus's sun-centered-universe theory came at a time when astronomers realized that earth-centered models of the universe could not adequately explain the movements of the planets.

2. *The Starry Messenger* (1610). This book reported Galileo's first telescopic observations. Two observations dealt lethal blows to the old earth-centered universe. First, Galileo observed that the planet Venus went through phases like those of the moon. This observation was impossible according to the geometrical models of the earth-centered universe. Yet this observation corresponded exactly to what one expected to see in a sun-centered universe. Second, Galileo observed the planets as spherical discs. Like the earth, they were round balls of matter. Planets thus became places, not objects or gods.

3. The *Principia* (1686). In the *Principia* Newton presented his famous law of gravity. This mathematical model removed the mysticism from planetary motion. Motions of the solar system now fell within the realm of human understanding. The law of gravity allowed astronomers to predict precisely where planets would be in the future. It even permitted astronomers to locate a previously unknown planet, Neptune.

4. *The General Theory of Relativity.* In 1916 Einstein finally answered a question that had perplexed many astronomers. If the earth did not occupy a central position in the universe, what object did? The Special Theory of Relativity denied the existence of any fixed point or any absolute space in the universe. Einstein abandoned the concept of absolute space in order to explain how light always travels at the same speed through a vacuum.

The General Theory of Relativity demonstrated that the presence of matter in the universe curves the geometry of space. Photons of light, as well as particles of matter, travel in a curved path rather than in a straight line. The entire universe possesses an overall curvature. Finding the "center" of this curved universe would be as meaningless as finding the "center" of the earth's surface.

Consider a presocratic Greek philosopher resurrected into the twentieth century. Imagine his astonishment when he finds that the capital of his old nation-state is not at the center of the earth; that the earth is not at the center of the solar system; that the solar system is not at the center of the Milky Way galaxy; that the Milky Way galaxy is not at the center of the Virgo cluster of galaxies; that the Virgo cluster of galaxies is not at the

center of the Abell supercluster of galaxies; that the . . . oh well, considering the General Theory of Relativity, there is no center to look for anyway.

In 1964 I visited the Griffith Observatory and viewed a planetarium program discussing man's place in the universe. By the end of the hour's presentation, the lecturer had thoroughly succeeded in flushing mankind out of the center of the universe. Not wanting to leave his audience dismayed with the earth's astronomical insignificance, the lecturer attempted to end the show on a cheerful philosophical note: "Considering all the vast galactic systems in our universe, man's habitation on one small planet seems rather unimportant. Yet it is the mind of man that comprehends the relationship between the earth and the other objects in the universe. Man alone possesses the ability to understand and explore worlds beyond his own. Granted, the earth may be small, but because of human intelligence, its significance overwhelms its size."

Our planetarium lecturer gave a simple reply to the astronomical decentralization of earth. He merely said, "So what?" Religions have replied with the same answer. After all, does the location of the earth really affect man's relationship to the universe? No, because if man has great significance in the universe, the significance comes out of *what* he is, not *where* he is. Yet, even after centuries of systematically flushing man from the center of the universe, astronomers have not yet undermined his intrinsic importance for what he is—a creature of limitless intellectual potential.

But recently astronomers have undertaken what might be called the "royal flush"—the ultimate decentralization of man. The royal flush denies man a central location in the universe and, even more important, it denies him any position of unique or supreme intelligence.

Recall that in 1610 Galileo published his telescopic observations in a book entitled *The Starry Messenger*. This treatise revealed that planets were not just moving objects, but real places resembling the earth. Predating Galileo's *Starry Messenger* by more than ten years, Giordano Bruno suggested that stars were places like our sun. He considered these stars to have planets inhabited by potentially intelligent creatures. Giordano Bruno challenged the scholars of his generation with the royal flush. He supported Galileo's displacement of the earth from the center of the solar system. He denied the conviction that intelligent life could exist only on earth. For continuing to teach these radical, for his time, ideas Bruno was convicted of heresy and burned at the stake.

Being burned at the stake tends to put a damper on even the most promising career. Bruno's example served notice to astronomers to avoid publicizing convictions that intelligent life existed on other planets. Thus, during the first three centuries following Galileo's use of the telescope, astronomical research did not strongly support a universe teeming with life-bearing planets. Up to the late 1920s astronomers knew of only one galaxy, our own Milky Way. Astronomers could not ascertain whether, in this Milky Way of some one hundred billion stars, our solar system resulted from a rare catastrophe or from a relatively common natural formation process.

Only during the past two decades have astronomers finally found that planets commonly form during star formation. Recent observations reveal that a high percentage of the nearest stars—stars within twenty light-years of the sun—have planetary systems. According to the research of Peter van de Kamp, Barnard's star, the second closest star to our sun, possesses two massive planets. One roughly equals the mass of Jupiter, while the other roughly equals twice the mass of Saturn. In addition to Barnard's star, 61 Cygni and Lalande 21185 also possess planetary systems. Very recently the list of nearby stars with planetary systems has grown to include Cin 2347, BD+43° 4305, 70 Ophiuchi A, BD+20° 2465, AOe 17415-6, Kruger 60A, BD + 5° 1668, Epsilon Eridani, and Proxima Centauri, the star closest to the sun. The high number of nearby planetary systems demonstrates that planets and stars form together as they both contract from clouds of interstellar gas. Planetary formation results from a common natural formation process.

Now consider the consequences of even one galaxy, our own Milky Way, filled with hundreds of millions of planetary systems. Does it not seem incredible, in fact, unbelievable, that only one of these systems should have intelligent life? Astronomers almost universally agree that the answer is yes: There must be other planetary systems in our galaxy that possess intelligent life. Astronomical observations now support mankind's ultimate decentralization—a denial of his being the only creature to possess intelligence as well as his removal from any special location in the universe.

Consider that man does not stand alone as an intelligent creature in the Milky Way galaxy. Logically, an important question follows. At this very moment, how many intelligent civilizations share the Milky Way with us? Answering this question requires that one assume that life processes elsewhere in the universe are consistent with those we find on earth. This does not imply that astronomers expect to find all intelligent life looking like human beings. Rather, it implies that all life in the universe, whether simple or complex, has a chemical foundation based upon long molecular chains of the carbon atom. Only carbon molecular chains seem to build substances that exhibit all of the five attributes associated with "life," that is, (1) reaction to stimuli, (2) reproduction, (3) respiration, (4) digestion, and (5) excretion.

Granted, there could be other substances, as yet not even discovered, that may exhibit these attributes of life. But consideration of these unknown life-forms would only allow astronomers to hazard a wild guess as to their frequency in our galaxy. By restricting life to the same chemical base that we find on earth, astronomers can restrict their search for life in our galaxy by looking for planetary systems resembling our own solar system. From a knowledge of how many such planetary systems exist within the Milky Way, astronomers can estimate the number of planets with earthlike conditions; and from this, the number that possess intelligent civilizations.

According to Carl Sagan, professor of astronomy at Cornell University, the number of communicating civilizations present in our galaxy today is a function of seven variables: (1) the mean rate of star formation over galactic history; (2) the fraction of stars with planetary systems; (3) the number of planets per planetary system with conditions ecologically suitable for the origin and evolution of life; (4) the fraction of suitable planets on which life originates and evolves to more complex forms; (5) the fraction of lifebearing planets with intelligence possessed of manipulative capabilities; (6) the fraction of planets with intelligence that develops a technological phase during which there is the capability for and interest in interstellar communication; and (7) the mean lifetime of a technological civilization. Let us now consider each of these seven variables in greater detail.

The mean rate of star formation throughout our galaxy's history tells us the average number of stars that form each year. Our galaxy contains approximately one hundred billion stars. These one hundred billion stars have formed over a ten-billion-year period. Dividing the number of stars by the number of years gives us a mean rate of star formation of ten per year. However, star formation undoubtedly occurred more rapidly early in the Milky Way's history. Today it occurs much more slowly.

Now, of these ten stars that form each year, how many possess planetary systems? Astronomical observations now show that planetary systems develop as a natural consequence of star formation. While large masses of gas contract to form stars, smaller gaseous masses condense to form planets surrounding these stars. Barnard's star, Lalande 21185, and 61 Cygni, all close neighbors to the sun, move through space in a deflected path explained by the gravitational pull of a nearby, massive, unseen Jupiter-like planet. This evidence of other planets compels astronomers to believe that our solar system could not have formed from a rare catastrophic event. Rather, the existence of other "solar systems" demonstrates that planetary formation must result from star formation. Since planets naturally form as stars form, nearly all stars possess planetary systems. Thus the fraction of stars with planetary systems approaches 100 percent, or unity.

With nearly every star possessing some system of orbiting planets, what fraction of these planetary systems have at least one planet with environmental conditions suitable for the origin and evolution of life? Fortunately planets do not necessarily have to be earthlike in order to develop life. In fact, when the earth first developed life-forms about three and a half billion years ago, its environmental conditions differed from those experienced by us today. The earth's atmosphere contained large amounts of ammonia, methane, helium, and hydrogen. These gases, mixed with large amounts of water vapor and water droplets, formed an atmosphere that resembles that of the planet Jupiter today. Not only does Jupiter's atmosphere resemble that of the primitive earth's, it also has earthlike temperatures. Slow gravitational contraction heats Jupiter's interior so that its outer layers of gas receive warmth from its own internal heat source. Jupiter would continue to have life-supporting conditions even if the sun burned out.

The nearby stars that are known to have planetary systems all have at least one planet with a mass approximately equal to that of Jupiter. Only the existence of a massive planet deflects a star's motion through space sufficiently to allow the planet to be detected indirectly. Thus, for all the known planetary systems, at least one planet in each system resembles Jupiter. Jupiter possesses ecological conditions that could support the origin and evolution of life. Jupiter generates life-supportive temperatures from its own internal heat source. Therefore, Jupiter-like planets elsewhere in the Milky Way almost certainly have life-supporting environments. Also, Jupiter-like planets probably exist in every planetary system. After all, our own solar system has three other planets closely resembling Jupiter—Saturn, Uranus, and Neptune. The abundance of Jupiter-like, potentially life-supporting planets allows astronomers to conclude that nearly all existing solar systems have at least one planet conducive to the development and evolution of life. Therefore the fraction of planetary systems with potentially life-supporting planets approaches 100 percent, or unity.

Of the potentially life-supporting planets, what fraction would permit life to evolve into complex forms? Remember that life neither marks the beginning nor the end of a complex cycle of chemical evolution. Biologists agree that the very conditions that allow life to develop also permit life to develop to greater complexity. So every planet that develops life, *given enough time,* should also develop complex forms of life. The fraction of life-supporting planets that develop complex life-forms again approaches 100 percent, or unity.

Of the planets that develop complex life-forms, what fraction develop life-forms with higher intelligence? The transition from complex life-forms to intelligent life-forms does not constitute a unique evolutionary process. Intelligence merely serves to increase a life-form's chances for survival. Biological evolution toward greater intelligence naturally occurs as different organisms compete for survival. As organisms increase their biological complexity, so will they also increase their intelligence. Given sufficient time, intelligent creatures should emerge on all planets with environments suitable for the development and evolution of life. The fraction of planets that possess intelligent life-forms approaches 100 percent, or unity.

Of the planets that possess intelligent life, how many will develop a technical civilization capable of interstellar communication? Optimistically, one may consider the development of technology a natural consequence of the evolution of creatures with great intelligence. But intelligence alone does not always result in the creation of a technical society. Dolphins, for instance, possess a greater brain weight in proportion to their total body weight than do humans. However, their adaptation to a liquid environment renders their bodies incapable of any manipulative functions. Unable to use any form of tool, dolphins will never build a technical society despite their intelligence.

In fact, if we consider only human life, we see that not all societies have developed technically oriented cultures. Just one culture—the European—has developed a scientific civilization. It has eventually dominated all other cultures by either destroying them or forcing them to adopt its technology. The dominance of this one culture has made impossible any estimate of the percentage of human cultures that might enter a highly technical phase. Astronomers cannot accurately know whether the earth's one technical society has been a fluke occurrence or a natural event among intelligent creatures.

Since progression from an intelligent society to a technical society may be partly a chance occurrence, let us assume that the fraction of planets with intelligent life that develop a technical, scientific society is 1 percent. This fraction may seem pessimistically low, but this low value should compensate for an optimistic estimate for the fraction of planets that develop intelligent life-forms.

Before evaluating the last variable—the mean lifetime for a technical civilization—consider again the variables just discussed.

R_*	Mean rate of star formation	10
f_p	Fraction of stars with planetary systems	1
n_e	Fraction of planetary systems with planets suitable for the origin and development of life	1
f_l	Fraction of suitable planets where life develops to complex forms	1
f_i	Fraction of planets with complex life-forms that develop intelligent life	1
f_c	Fraction of planets with intelligent life that develop technical civilizations capable of inter-stellar communication	1/100
L	Mean lifetime of a technical civilization	?

The number of technical societies in our galaxy capable of interstellar communication equals the product of the probability of occurrence of each variable. According to Sagan, the mathematical formula, where N equals the number of technical, scientific civilizations existing within our Milky Way at this moment in time, is:

$$N = R_* f_p n_e f_l f_i f_c L$$

Now let us solve the equation using the values we have established for each variable.

$$N = 10 \times 1 \times 1 \times 1 \times 1 \times 1/100 \times L$$
$$N = 1/10 \times L$$

The number of technical societies in the Milky Way equals one-tenth their average lifetime. Thus the number of technical societies currently existing in our galaxy depends completely on the figure given for the least-known variable—the average lifetime of a technical society.

The earth has possessed a technical civilization capable of interstellar communication by radio transmission and reception for about twenty years. Yet our technology could end at any moment. The problems of overpopulation, pollution, and nuclear war threaten the existence of mankind. Newspaper headlines daily remind us of the possibility that humanity could perish within thirty years. If this pessimistic outlook comes true, the lifetime of our technical society would equal just fifty years. If we consider our society typical of all other technical societies, then the value of L in our equation should equal fifty.

$$N = 1/10 \times L$$

With L equal to fifty years, the number of technical societies currently existing in the Milky Way equals 1/10 x 50, or just five! Five civilizations scattered randomly throughout our Milky Way would be separated by an average distance of over twenty thousand light-years. If we on earth decided to say hello to someone twenty thousand light-years away, it would take our radio message twenty thousand years to reach its destination. Their reply would, of course, take twenty thousand additional years to reach us. In other words, this instance of communication would take forty thousand years. Obviously such distances render interstellar radio communication essentially impossible. If one accepts the pessimistic view that all technical civilizations extinguish themselves within about fifty years, then one must deny the possibility of interstellar communication. Not only would the distances between civilizations be far too great, but their limited time for technological growth would not allow for the development of communication channels other than radio transmission.

But let's adopt a more optimistic and, we hope, a more realistic attitude toward the lifetime of technical societies. Assume that the proposed fifty-year lifetime for the earth's technological society is typical for *most* but not *all* scientific civilizations. Consider that just 1 percent of emergent technical societies pass through their population-pollution-nuclear crisis. They then proceed to advance technologically, free from destructive wars and conflicts. Under this climate of peaceful progress, these scientific civilizations endure for an unlimited number of years.

Let us now compute the value of L (the average lifetime for a technological civilization) where 99 percent of technological civilizations endure for fifty years and where 1 percent of them endure for five billion years. Under this assumption the mean lifetime for a technical society would be fifty million years. Plugging the figure fifty million for L into our formula—$N = 1/10 \times L$ (50,000,000)—we find that five million technologically advanced civilizations exist in the Milky Way today. Thus if just one in a hundred scientific societies survives its own technology, then millions of them currently exist within our galaxy. With millions of these societies existing in our galaxy, their average separation would be only one hundred light-years. With so many technical societies nearby, one begins to wonder why we have not already received word from them. Remember, though, that daytime television emits the strongest radio signals from earth to outer space. For years upon years nearby civilizations have exclusively listened to "The Lucy Show" and "As the World Turns." It is no wonder they may have carefully avoided communication with earth.

Many individuals, however, claim frequent communication between the earth and other planetary systems. This communication does not result from the exchange of radio signals but rather from actual visitations by interstellar spacecraft. Frequently astronomers receive reports of either sightings or landings of extraterrestrial spacecraft. In addition to these current sightings, certain authors claim that many ancient drawings and temples imply contact with extraterrestrial beings. Yet most of these UFO reports, whether ancient or modern, share one remarkable commonality. They picture these spacecraft as either resembling current military rockets or experimental lifting-body aircraft (flying saucers).

Consider these UFO observations in the perspective of our knowledge of the mean lifetime of advanced civilizations. With an average lifetime of fifty million years a typical extraterrestrial visitor, whoever that might be, would possess scientific technology millions of years in advance of our own. He would have already exceeded our own technology at the time the Eohippus, the first horse, walked the earth. Contemplate such an advanced technology. Would they travel across vast stellar distances in vehicles already used by or conceived of by the United States Air Force? Not likely. It is far more reasonable to imagine that these creatures would beam down to earth a la "Star Trek" than that they would land in some type of Flash Gordon rocket. Most UFO reports actually contradict the one thing that we can assume about extraterrestrial visitors—their incredibly sophisticated scientific technology.

Remember that scientific technology grows exponentially. More technological growth occurred between World War II and 1970 than in all the years of recorded history before that time. The decade of the 1970s will see more scientific progress than there was during the years between World War II and 1970. The rapid pace of technological development does not require that a civilization be millions of years ahead of us to vastly surpass our scientific knowledge. Indeed, a technical society just a century ahead of our own would be as advanced from us scientifically as we are from the Roman Empire.

Realize now that our earth has just entered a technical phase capable of interstellar communication. Providing that scientific societies survive their own technology, we have just entered a galactic system of communicating technologies. Since these technologies exist millions of years in advance of our own, we must realize: they are relatively intelligent, and we are relatively stupid.

For over half a century earth radio communications have traveled outward through interstellar space. These waves now touch planetary systems fifty light-years away. Already a reply to these radio emissions could be traveling earthward. At any time, radio telescopes could reveal to all the world that human intelligence does not reign supreme. Human wisdom—the youngest, most immature mind within a galactic society of millions—has just received membership in that society.

When thinking about ETIs—about which we know absolutely nothing, of course, and therefore can speculate to our hearts' and minds' content—we are prone to make assumptions about intelligence derived solely from our limited terrestrial experience. But in biocosmic thinking we are pushed to transcend this limitation. What exactly can we validly assume about *intelligence* per se?

Although Peter Angeles is also a poet and playwright, his philosophic credentials rest primarily on his skills in logical analysis; and in this article for *ETI* he goes to work on the assumptions and implications of the "biocosmic belief."

Is the biocosmic world view of philosophic importance for our time? "Enormously important. But in some ways it's scary. ETIs may really be far, far beyond us, and we may feel insignificant by comparison. On the other hand, we just may feel a new kind of challenge to improve our capacities, to expand our mental outreach as never before."

Angeles has written *The Possible Dream* (1970), *The Problem of God* (1974), *An Introduction to Sentential Logic* (1976), and has two books forthcoming: *Critiques of God* and *An Introduction to Categorical Logic.*

Dr. Angeles teaches philosophy in and around Santa Barbara City College in California.

"If extraterrestrial intelligence is never found at any time in the universe, the quest will still go on: there will always remain a possibility in an infinite universe that extraterrestrial intelligence exists somewhere, *or did exist* somewhere, *or will come into existence* somewhere. *In an infinite universe the biocosmic belief cannot be eradicated."*

Peter Angeles

ETIs and
the Problem of Intelligence

We are not alone. Life is a universal process. Life-forms—high life, low life, all sorts of life—are to be found "out there." Life exists in all imaginable, and in countless unimaginable, configurations. Life is neither an earth accident nor a "finger-of-god" disturbance. It cannot be unique to our blue-green planet home. How will acceptance of this fact affect our thought and feeling? —James L. Christian

I

There are three initial dimensions to the biocosmic belief: (1) an actual *physical* confrontation and attempt at communication with extraterrestrial intelligences; (2) a successful communication and exchange of information with extraterrestrial intelligences (but not a physical confrontation); (3) neither a physical confrontation nor any communication, but the *quest* for contact with extraterrestrial intelligences, and all that is entailed by that quest.

161

If extraterrestrial intelligence is never found at any time in the universe, the quest will still go on: there will always remain a possibility in an infinite universe that extraterrestrial intelligence exists *somewhere*, or did exist *somewhere*, or will come into existence *somewhere*. In an infinite universe the biocosmic belief cannot be eradicated. The lack of evidence for extraterrestrial intelligence at any point in an infinite time span would not logically entail that there is no evidence to be had. In the biocosmic belief it would merely signify that no evidence was yet available but that evidence could be had.

Complex organic compounds and, perhaps, extraterrestrial life-forms lower than man will be found in our solar system. Probably these will be found much earlier than intelligences in other solar systems (that is, extrasolar intelligences). There is no way at present for us to detect complex organic compounds, or nonintelligent life-forms, in other solar systems. A major problem is to detect planets around other suns. If intelligent life-forms are found in other solar systems, they will probably be found when they manifest an intense intellectual activity trying to communicate with us by such means as radio or laser communication (or by direct contact, or visitation). That communication, if it ever happens, could of course be decoded earlier than we could find complex organic compounds and nonintelligent (lower than man) extraterrestrial life-forms in our own solar system. But it seems safe to say that for some time to come we will not be able to detect complex extrasolar organic compounds, or nonintelligent life-forms outside our solar system, by direct empirical evidence such as the analysis of soil, plants, or organisms brought back by spacecraft.

The biocosmic belief would insist that the greater the number of extraterrestrial life-forms found in our solar system, the stronger the presumptive claim for there being higher life-forms up to (and beyond) the intelligence of man in other solar systems. The central philosophic problem for the biocosmic belief will then shift from the question, What is life? to, What is intelligence? How do we recognize that something is exhibiting intelligence, especially if its structure is grossly unlike that of the life-forms around us that we recognize as displaying some degree of intelligence? A true biocosmos will deal not only with a higher form of intelligence but will also have to deal with higher forms of intelligences—and intelligences foreign in structure and function to each other. (Stars vary in age as well as in size. In all probability biological evolution occurs at unequal rates on planets in other solar systems. Among other things, this would mean that evolutionary development is different from solar system to solar system.) How will communication be possible and what form might it take? What would be the likely repercussions of relating to higher intelligent forms? The ultimate question may not be one of how we should relate and what we should do but of how we can survive.

The most profound changes will come about when we actually find and confront extraterrestrial intelligence, or are confronted by it. But this actual physical contact with extraterrestrial intelligences need not occur in order for us to experience profound changes in our modes of behavior—morals, laws, philosophy, arts, religion, science, technology. Let there be merely a communication tie-up with extraterrestrial intelligences and the subsequent exchange of technological information will produce those changes. Even if *no* extraterrestrial intelligence is found and *no* communication ever established, the very search for extraterrestrial intelligence, if taken seriously and undertaken with intensity, will affect our civilization in ways that few, if any, other quests of man have. For the biocosmic belief man's future will be altered simply by the desire and quest for contact with extraterrestrial intelligences. (Could an extraterrestrial intelligence be so advanced and so different from anything in our awareness that we cannot detect them but they can detect us? Could it be that they do not want to communicate their recognition of us to us and that, with their complex technological knowledge, they are capable of eventually controlling the direction of our history and species without our knowing it?)

Our questions about extraterrestrial intelligences will be a cycle somewhat like this: Do they exist? How do we know? Where are they? What are they like? "Who" are they? How many are there? What can they do to us? What can we do to them? What are they doing to us? (Help!)

Our children and our children's children are entering an era where not only must they adapt to rapidly accelerating and astonishing technological, scientific, and social changes, but where the time lapse between some of our most unbelievable science fiction and its realization will be only months and years rather than decades and centuries.

Intelligences no more technologically advanced than we are could detect and send—as we are able to do—coded radio transmissions over distances of a thousand light-years—an area in which there are ten million stars. There is a high degree of probability that in that vast number of stars there are at least a few planets that do have the capability and interest in attempting radio communication within an area of one thousand light-years around them and are doing so. It would take mes-

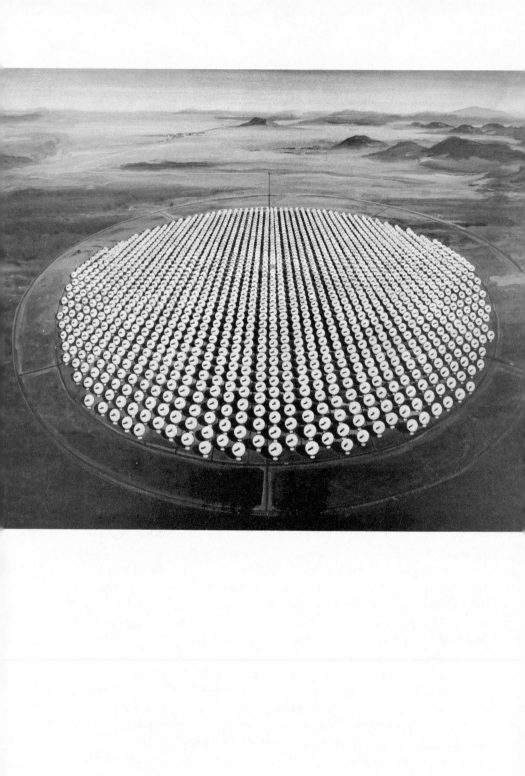

sages within that radius up to one thousand years to get here, and our reply would take as long to get there. The content and purposes of our communication would be drastically different from the types of communication we use today. It would not be communication between individuals (who would long be gone from the scene) but communication of a long-term programmed bank of stored questions and answers devised by institutions that have been assigned the task of continuing an uninterrupted stream of communication (such as answering thousand-year-old questions possibly by having to extrapolate what the answer would be a thousand years hence, so that when the answer got back it would be somewhat current).

II

We might also look at some of the assumptions that I believe a biocosmic belief would support. Most of these assumptions must be seen in tension with their possible opposites.

1. *Man's intelligence is not unique in the biocosmos.* (Opposite: Man's intelligence is unique in the biocosmos.) The origin and manifestation of man's intelligence may be regarded as unique only in the sense that it has a specific and particular evolutionary history within a universal life-creating process. Man's present state of intelligence has existed, does exist, and will exist repeatedly elsewhere in the biocosmos.

2. *Man's intelligence is an average intelligence among extraterrestrial intelligences.* (Opposite: Man's intelligence cannot be said to be an average intelligence among extraterrestrial intelligences. We cannot talk about "average intelligence" since there are no samples of other intelligences from which to take an average. We have not encountered any other intelligences yet, and there may not be any other intelligences higher than ours anywhere in the universe—vast as it is!) Given the fantastic number of possible planets on which intelligences can develop, the likelihood is that man's intelligence as it now exists is an average intelligence.

3. *There are different kinds of intelligences existing in the biocosmos.* (Opposite: There are no different kinds of intelligences existing in the biocosmos. There is so far not even any proof of the existence of any other intelligence than man's. And if other intelligences were found and they are different, then to that extent they could not be called "intelligences." "Variety of intelligences" is a contradiction in terms.) There are varieties of life functions within the class of life structures. There are varieties of levels of development of these life functions within the class of life

structures. There are varieties of levels of development of these life functions and life structures.

The same, or similar, life structures do not have to exist in order for "intelligent" (that is, "intelligible") communication to be detected or even created. This is the case especially if there is a property or method common to a procedure, or interaction with a specific type of relatedness, or response with a systematic pattern of interactions. These are relational processes (our language tends to give them the substantive connotations we give such words as *properties, forms, patterns*) common to "intellectual" content of any kind, irrespective of the origin of that "intellectual" content and irrespective of the nature of the structure "having" it or "creating" it. These common relational processes are necessary to the solution of certain kinds of problems or to the response made to certain kinds of impinging interrelatednesses. The analysis of the components of these relational processes will have no reference to human content as "human," in the sense of any peculiar inner feelings and emotions of the source. The analysis will be essentially contentless and thereby highly mathematical. Programs designed to communicate our real-inner-world on the basis of our natural language will be inadequate to the task. A more formal "language" needs to be developed for extraterrestrial communication, one capable of expressing the most abstract and general relational processes applicable to human beings and to mankind. (Some aspects of this will be touched on in section three.)

4. *The present state of man's intelligence has been surpassed and will be surpassed.* (Opposite: Man's present state of intelligence has not been surpassed.) There are stages of development of intelligences so far superior to man's that they are as undecipherable to man as man's technology is undecipherable to his pets.

5. *The present state of man's intelligence is not a final product of the biocosmos.* (Opposite: Man's intelligence is a final product of the universe and is not evolving toward greater complexity.) There are no final products in a biocosmos, except the immortality of life functions as such. There are no final goals or ends "intentioned" by a biocosmos. There is only an infinity of ongoing events in a plethora of stages interrelating and disintegrating, succession upon succession.

Man's science and technology can provide the impetus toward developing the present state of man's intelligence into completely new, and at present unpredictable, complex and advanced functions. Whether this happens or not may depend as much on man's politics as it does on natural or historical accidents. Whether it happens is not of paramount importance to a biocosmos where there are the seeds of an infinite number of other such opportunities. If it matters at all, it will have to matter to *us*. And if it doesn't matter to us it will not matter at all. Man has arrived at a position in the biocosmic life-creating process at which, if he does not surpass his present self, he will be surpassed or become extinct.

6. *The sophisticated technological advance of man's intelligence and the necessary adaptation to his most pressing crises are linked to man's concentrated technological effort to make contact with extraterrestrial intelligences.* (Opposite: The sophisticated technological advance of man's intelligence and the necessary adaptation to his most pressing crises are not linked to man's concentrated technological effort to make contact with extraterrestrial intelligences. That would be a waste of time, money, and effort. Rather, it depends upon man's concentration on improving himself as a rational and humane creature and on improving his social structures to allow the actualization of his potential and to avoid what appear to be the imminent crises he is facing.) There are stronger versions of this assumption: (a) In order to achieve the necessary advance of his intelligence for survival (survival of his own globe's problems and of the possible challenges of extraterrestrial visitations), man must search for and prepare himself to find evidence of the existence of extraterrestrial intelligences. (b) In order to achieve the necessary advance of his intelligence for survival man must gain information from communication with more advanced extraterrestrial intelligences, from whom he can learn, or by whom he can be challenged to learn or who can help provide some of the solutions to man's rapidly increasing and seemingly insurmountable problems preventing him from progressing to a higher state of intelligence. Even if the civilizations of these other intelligences did not face and solve the same social and technological problems that we are and will be facing, they would still be able to creatively critique our approaches to solutions—provided they are sympathetic to our plight and we cannot do it on our own (as seems increasingly to be the case).

Success in communicating with extraterrestrial intelligences means that we can gain invaluable information on higher stages of biological and social evolution—what they are like, how they got to that stage. We will be able to determine the extent to which we are (or are not) heading in that direction, what we can do to imitate (or avoid) that process of advance, and whether it is in any way applicable to us. This would be seeing possibilities for our future and predetermining which route we would like to take.

To survive, man must become a more future-looking and future-controlling creature. Little planning for today was done by our forefathers fifty, one hundred, two hundred years ago. We cannot allow ourselves and our descendants to be uninterested in predicting, planning, and preparing for the future use of our resources and the development of mankind. Getting data from extraterrestrial civilizations would be extremely relevant and perhaps at some point absolutely necessary. I am not suggesting, however, that we should stop our own meager efforts toward improvement and wait for information on biocosmic civilizations to solve our problems.

7. *Man's intelligence is one expression of an inescapable tendency for interrelatedness in the biocosmos. There is, of course, the counter-thrust of disintegration.* (Opposite: Man's intelligence is a manifestation of an interrelatedness of events in nature, but there is no "inescapable tendency" in man's intelligence for interrelatedness with other biocosmic phenomena.) *Intelligence* is merely a word. It may be used to name or characterize a great number of things—with no complete agreement or precision as to the characteristics to which it is to be applied. Cultures seem to have metaphors to help them understand what they are trying to understand. Accepting the concept of programming as our current leading metaphor and accepting its application as a loose analogy, it is possible to see that all "intelligence" has a programmed aspect to it. "Higher intelligence" has the "program" aspect of being able to discover, to invent, and to extend new "programs" found in its interactions and attempts at interrelatedness. All "intelligence" is programmed to become interrelated and to enter into the future in a certain way. (This view allows the application of the word *intelligence* to such things as "self"-replicating systems.)

"Higher intelligence" is characterized by the quest for information (and thereby its search for interrelatedness) about (a) what is known about the world, (b) how it is known, (c) how this "known" can be represented and/or replicated, and (d) what *can* be known, that is, in what forms or within what parameters. The third question is an "entering into the future"—a continuation of an existent in nature for a longer time than would otherwise be the case, or, so to speak, a short-term immortality of that existent. These are manifestations on a higher level of a thrust, a tendency (animism!) in the universe to retain into the future something of its past presence. Thus at any given time in an eternal universe there are existents that have (for some time) some similarity to existents of the past. (There is *no* time in the universe at which something in it resembles nothing else.) "Intelligence" is that aspect of the universe in a "duration" state, entering into the future in a certain way, based on the incorporation of its past and the integration of its present (its "programs"). "Higher intelligence" has the added complexity of being that aspect of the universe in a "memory" state (its own "duration" state), entering the future in a certain direction based on its analysis and understanding of the past and its integration of the present.

There are common features to similar-order intelligences. (I am suggesting that there are also features common to *all* orders of "intelligence.") These features are not only common but also objective, that is, *not* dependent on the peculiarities of the structures of the originating source. Intelligence as exhibited in a function is a process independent of the particular chemical, biological, or cultural genesis. What these common and objective features are will be determined not by the nature of the source of that intelligence but by such things as the description of the abstract relational processes being displayed, the nature of the quest or of the problem being pursued, and the patterns of activities and interrelatednesses entered into. We should expect to find in the universe other similar-order intelligences using common relational procedures for resolving problems, for organizing systems procedures, for interrelating "contentless" relational patterns. Extraterrestrial intelligences need not be of the same structures as ourselves in order to have intelligence. If they are of our order intelligence, or slightly higher, then they will desire to communicate; and if they attempt to communicate, they clearly have intelligence.

The actual exchange of information (or conflict) among extraterrestrial intelligences in this drive for interrelatedness will produce changes with more far-reaching cosmic effects than even the existing vast number of atoms on their own would be able to produce. A "natural" property of some intelligences is to make their presence detectable at long range or to desire to make physical contact. We will enter the realm of cosmic history, a history lived and written on a cosmic time-scale, with biocosmic processes—not mankind—as the leading character.

III

The question of extraterrestrial life is perhaps a distinct question from the question of extraterrestrial intelligence. I say "perhaps" because there is the remote possibility that characteristics of "intelligence" could exist in the universe as properties of nonliving processes. I have no idea what that might be like. I merely note it as a possibility. Nor do I have a definite idea of what constitutes "intelligence." The more specific I become about the meaning of the word *intelligence*, the more I find myself limiting its application to quasi-human traits; the more general I become about its meaning, the more I find that it applies to most things in the universe, even molecules and atoms. Does intelligence—however it is defined—have to be associated with a neurologic-type organism, or are there other forms of expression of intelligence far different from this neurologic type? (This may not be a pressing issue to most people. We have problems enough deciphering the "intelligent" activities of living forms around us, such as dolphins. Also, in an infinite universe there are bound to be "intelligences" exhibited in *living* forms to occupy our attention.)

My answer to the above question has already been formulated in general: intelligence can have its source in phenomena quite different from the biochemical and neurological processes found in organisms—there are other forms of expression and sources of intelligence different from these. Here is a simplified example providing an extremely general awareness of this possibility. Computer-programmed machines may not be said to have motives, or motivational desires. Yet, when certain expressions in the computer's "memory" are given the identification of being "goals," we can interpret this activity as "motivational" even though it is a machine (without "motives"). The computer program "contrasts" a present situation with the expression identified as a "goal." Those expressions leading to the goal, or generated to get to the goal, are each in turn subgoals, which are "pursued" to get to the goal. The computer program, in drawing contrasts and comparisons, in generating the subgoals, is, in its activity, drawing up possible courses of action (possible subgoals) to get to the goal, and as it goes along it is "predicting" possible results and comparing them with the goal to see how close it is to the goal and what remaining avenues there are to reach the goal. These programs have most of the *functional* manifestations of the meaning of motivation, but their *structures* originating this activity are far different from the human (neurological) structures producing human motivation.

If extraterrestrial intelligences exist, we would have to create some system or language for communicating to them some of the things we are "intelligent" about and then wait for some response from them to our system. Only when we receive a response can we say that extraterrestrial intelligence has been discovered. This is easier said than done. In the vast expanse of the universe it would require centuries for the exchange of messages, even if the exchange occurred at the speed of light. The universe's infinitely vast distances contain no split seconds for our consciousness—nothing happens instantly relative to those dimensions. Extraterrestrial intelligences would have to determine: (1) that our signals are attempts at communication and not natural phenomena, (2) what repetitive contents and patterns are being used as the basis for the communication, (3) what these patterns "mean" or refer to, (4) how to indicate that they have been received (send them back just as they were received!) and are being answered, and (5) how to use this system to communicate further meanings and referents.

Detection systems for extraterrestrial communication signals have been devised. (Even the slightest indication of a successful scientific decoding of a message being broadcast by a civilization hundreds of light years away from an infinitesimally small spot out in the universe unreachable by man would be enough to change the behavior and research of our entire planet.) Codes have been constructed and sent as messages to be deciphered by extraterrestrial intelligences. The codes will become even more ingenious as time goes on. Codes will be invented that will be decipherable by extraterrestrial intelligences with structural make-ups far different from our own. We will be able, during an extensive interchange between us and these extraterrestrial intelligences, to arrive at a standardized, though highly mathematical, language that will be a useful tool for the communication of a limited range of our knowledge. That limited range of our knowledge that we will be able to communicate to each other in a special formal language will become technologically more and more important, useful, and crucial for the development and advance of our civilization if we are to surmount the technical problems for our survival, and perhaps for *their* survival. Nevertheless, other features of our knowledge, such as immediate feelings and experiences, will not be communicable—although that will not necessarily make a major difference—*unless* the participating intelligences have fundamental experiences close enough to our own to serve as foundations for a standardized language for communicating those experiences. Having the same fundamental experiences, or "knowledge by acquaintance," requires having similar receptors, that is, being similar living organisms. If the chemical, and hence structural, base for extraterrestrial organisms possessing intelligence is different, then exchange of sensory-type knowledge would be impossible (though exchange of other types of knowledge would be possible).

But doesn't this "subjectivistic predicament" apply as well to the mathematical-type, to this standardized scientific-type language that will be developed in communication with extraterrestrial intelligences? The subjectivistic predicament cannot be entirely avoided, but if its impact is to be minimized, then the standardized communication system will have to have a natural constant, or a constant we put into its signals, as its language base. This constant must be recognizable by extraterrestrial intelligences even with marked differences in structures and receptors, and even with markedly different mathematical notations. The external referent that will be the constant communicated will be the focus of attention and not the subjective states with which the subjectivistic predicament is primarily concerned. The contrast is between having sensations by means of particular receptors (and being able to have those sensations only by means of those particular receptors) and becoming aware of a pattern of constants being communicated. Such awareness can be produced by different structures—the particular ordering of the constants may be grasped, but the sensory state, if any, associated with that ordering may differ, or be nonexistent. The standardized communication system will have to communicate the common recognition of a natural constant (or a highly abstract constant we construct for recognition) as its language base. The important ingredient in extraterrestrial exchanges would not simply be that the constancy of relational patterns being communicated is known, but *that* it is recognized and in some way responded to.

But then what we communicate not only would be limited to an exchange of mathematically related and relatable recognizable patterns but also would be worthwhile only in a technological way, and no communication of our emotions, feelings, values, attitudes—in general of our psyche—would be possible. I am not unaware of the possibility that this limitation may be for the best. Nor am I unaware that this may be the direction our technology and precision communication is taking us—into a future where human problems and conflicts will have to be suppressed or eliminated since there are no solutions for them, but only adjustments that for the most part are difficult to maintain unless complete programming of human nature is achieved.

The following propositions have been assumed as working hypotheses in philosophy: (1) Similar structures produce similar functions—for example, similar brain structures of organisms indicate similarities of function, such as consciousness, memory, imagination, and sensation. (2) Different structures produce different functions. If, then, differences in functions are found, this has been attributed to differences in, or alterations of, the structure producing those functions. (3) Different structures cannot produce the same function. Thus it has been assumed that (a) the greater the similarity between structures, the more nearly identical are their functions, and (b) the greater the difference in structures, the greater the difference in their functions. With future advances in simulating "intelligent" activity in complex computer systems, we may find that propositions 3 and b may not be credible hypotheses. We may come to the conclusion that different structures can in some cases produce similar functions (for example, brains producing memory functions and computers producing memory functions—although they have different "structures").

If we could show that (totally) different structures existing as extraterrestrial intelligences could recognize (respond to) the *same* constants, would we thereby be shaking the foundations of the principle of relativity of (sense) perception? If we could prove that the same constants have been recognized by different systems of understanding, would this support the contention that some things can be known to exist independently of what the structures of our understanding do to them? Would we have to admit the realization of some "objective" realm understood apart from our unique methods of apprehension? Or would we insist that insofar as constants are recognized as the same, then similar systems of understanding are being used—that insofar as any constants are known in common, they can only be known, and must be known, relative to the similar categories of that system of understanding? Would it be argued that the *structures* are different but the *functions* (the understanding) similar and hence that the principle of the relativity of knowledge is still preserved?

There are two separate questions involved in this matter: (1) How do we know that extraterrestrial intelligences are seeing the same constant? This could be determined by such methods as establishing consistent responses from them to those constants. (Tautologically it might be argued that the extent to which they do respond to the same constants indicates the extent to which they have structures similar to ours for such apprehension; and to the extent that they have structures similar to ours, they will respond to the same constants. Would, one might ask, the empirical finding of extraterrestrial intelligences with [totally] different structures help us overcome the tautological nature of this problem?) (2) Are the images, thought processes, and concept formations taking place in different extraterrestrial intelligences similar to each other and to *our* images, thought processes, and concept formations?

This question can never be answered in terms of direct knowledge because of the unique privacy, the egocentric quality, characteristic of any unity of intelligence. It is not even in the same category as inferential knowledge based on observational or immediate knowledge. For example, let us say my neighbor and I look at the same patch of red. We have both been checked for color-blindness, and neither of us is color-blind. Is the image of red that my neighbor perceives similar to the image of red that I perceive? There is no method for arriving at a conclusive answer to this question. The fact that we *respond* similarly is not proof that our images of that red patch are similar. The response may be behaviorally and linguistically similar, but the "mind" realities may be different. I may try surgically to connect my brain to my neighbor's to see if our images of red are similar. But this would give only an image of the red patch that was a product of our *two* brains or *my* image of *his* image of the red patch— still a slight distance away from determining the similarity of *my* image to *his*. We can never get at his image separate from our image of his image. My image is all that I can get at because my image is all that I can have. Our own "mind" reality is all we can know and can hope to know since we cannot escape it to know anything that is other than it. Even if we did "escape" it, in what sense have *we* escaped?

The unfamiliar is known in terms of the familiar, the unknown in terms of the known. Together with the principles inherent in the egocentric predicament and the relativity of our realities, this principle of epistemological relativity appears to always hold. It has its problems: How is "new" knowledge gained? If it is the novel reorganization of present knowledge, does that imply that we potentially know all that there is to be known—a kind of Platonic doctrine of reminiscence? Or is all this simply based on a bad argument that we know what we know, and do not know what we do not know, and cannot know what we cannot know, and thus we know all that there is *to be* known—at least potentially?

My purpose here is of another sort. If it is true that we know the unfamiliar in terms of the familiar, then suppose something existed in the universe that was in all respects *totally* different from anything with which we are familiar—different from any form and content of our experience that we had ever had (or could have). Would we be able to recognize it? Do such realms or things exist in the universe? If the answer to these kinds of questions is yes, then our set of problems will be similar to the question: Have we somehow got the key for unlocking all there is to know, the true method for knowing the essential aspects of reality? But would we then be falling into the aristocentric trap of believing that what we presently know is the basis for all that there is to know and that from now on knowing what the universe is like will be just a continuation of our present structures of knowledge? The Babylonians, the Greeks, the Newtonians thought like that. What further realizations of kinds of reality are in store for us? Would there be some that could *never* be recognized? (And, of course, how would we know this?)

Suppose the answers to our questions are negative. Suppose that there are things, and that there always will be things, in the universe that we cannot recognize as existing. Our philosophical question would be: How do we, or how can we, ever know of such a thing, or even know that it exists? Our practical question would be: If we know that it exists, then we know of its existence; if we do not know that it exists, then we don't know that it exists. But what difference does it make? In one sense a difference is not a difference unless it makes a difference; and if it does not make a difference and we aren't even aware of any difference, then what is the problem? But this may be the crucial issue. Can something make a difference (by making a difference) without our recognizing it? You may not know that a thing exists unless you know in what respects it makes a difference, but something may exist that you know nothing about which is making a difference—a difference that itself is unrecognized (or in itself *unrecognizable?*). In biocosmic history it will become a real possibility that there are or can be extraterrestrial intelligences that are intentionally directing, or can intentionally direct, our behavior toward activities that can in no way be said to be due to *our* deliberations. We will *feel*—that is, be *made* to feel, without our knowing it—that we are in charge of our destiny; but in actuality we are being programmed and used as means for their short-term or long-term purposes.

Joseph Royce's professional life has been devoted to the proposition that individuals are "encapsulated" within a personal (partly genetic) way of knowing—a limited epistemic style or world view. Each of us is rational *or* empirical *or* metaphorical when, experiments show, we could de-encapsulate ourselves by entering more fully into all three ways of knowing.

Royce once wrote that "the key to moving toward relative unencapsulation, and thereby a deeper and more comprehensive world view, is to become more open to alien ways of knowing." By "alien" he meant the less congenial cognitive styles for each individual.

But in this article for *ETI* Royce speaks of "alien knowledge" in its full sense. There may exist truly alien ways of knowing that would have a revolutionary impact on our all-too-human, epistemic encapsulation.

Joseph Royce is a psychologist, interdisciplinary scholar, and philosopher. He is the author of *The Encapsulated Man: An Interdisciplinary Essay on the Search for Meaning* and *Inquiries into a Psychological Theory of Knowledge* (with Herman Tennessen); he is the editor of *Psychology and Symbol: An Interdisciplinary Symposium* and *Toward Unification in Psychology.*

Dr. Royce is founder and Director of the Center for Advanced Study in Theoretical Psychology at the University of Alberta.

"Extraterrestrial beings may well have a different conception of reality. . . . The existence of extraterrestrial intelligence would have an enormous de-encapsulating impact on man's world view."

Joseph Royce

Consciousness and the Cosmos

The key characteristic of man is that he epitomizes consciousness. This is evident phylogenetically if we contrast the discriminabilities of the amoeba with those of man. Similarly, the contrast in discriminabilities between the human fetus and the mature adult points to the ontogenetic basis of increased awareness. And the contrast between the world view of an aboriginal medicine man and that of Albert Einstein constitutes a dramatic example of the cultural evolution of consciousness.

However, in spite of man's impressive position on the awareness continuum, the more he learns of the cosmos, the less secure he is concerning the adequacy of his world view. In short, what remains to be learned seems to increase exponentially as knowledge increases. Each finding, while adding to the current stockpile of knowledge, opens the door to new questions and new mysteries. It is my guess that the human condition will always be this way. Thus, according to this view, all we can hope for is a world view or an image of reality, not reality itself.

While I personally do not agree with those who regard such a state of affairs as distressing, I am concerned about the relative narrowness of the typical world view that emerges. I have dealt with this problem under the rubric of encapsulation, my theory being that men tend to view the world through epistemically limited spectacles.[1] The point is that, although there is a multiplicity of valid approaches available to the reality seeker, individuals tend to highlight one of these approaches at the expense of the others.

The potential value of this book lies in its demand that we think in terms of cosmic consciousness. For the purposes of the theoretical speculations of this essay, I frankly do not think it matters whether or not there actually is extraterrestrial life. While an affirmative answer would undoubtedly change certain details, I doubt whether it would change the essay's basic thrust. The point is that the degree of variation in extant lifeforms on planet earth is so great that additional extraterrestrial variation would not change the basic question.*

And what is the basic question? The question is, How might highly intelligent extraterrestrial life-forms view the cosmos? Extraterrestrial beings, assuming they exist and have a culture, may well have a different conception of reality. Thus the crucial issue is not whether there is extraterrestrial life, but rather, how do we account for differences in world views.

EPISTEMIC STYLES, INDIVIDUALITY, AND WORLD VIEW

Elsewhere I have defined *world view* as "an organism's organised set of personal cognitions which constitute a model or image of reality."[2] My explanation for the existence of different world views involves a synthesis of the relevant segments of a general theory of individual differences[3] and the relevant segments of a psychological theory of knowledge.[4] The most relevant segment of the knowing theory says there are three basic ways of knowing: empiricism, rationalism, and metaphorism. These three isms are basic because of their dependence upon various cognitive processes, on

---

*I have assumed, of course, that possible extraterrestrial life-forms will have evolved in accordance with genetic-evolutionary theory as espoused by earthmen. However, even if extraterrestrial evolution is different (e.g., if it occurs by means of genetic engineering or the inheritance of acquired characteristics), the multidimensional model herein espoused should hold up. However, such matters as speed of evolutionary development would clearly be affected.

one hand, and their epistemological justifiability, on the other. This view can be briefly summarized by reference to Figure 1.

The implication here is that each of these isms represents a legitimate approach to reality but that different criteria for knowing are involved. Rationalism, for example, is viewed as being primarily dependent upon logical consistency. That is, it says that something will be accepted as true if it is logically consistent and rejected as false if it is illogical. Empiricism says that we know to the extent we perceive correctly; and metaphorism says that knowledge is dependent upon the degree to which symbolic cognitions lead to universal rather than idiosyncratic awareness.

While each of these cognitive processes may lead to error, the implication is that each is also capable of leading to truth. The possibility of perceptual error, for example, is readily apparent. The errors of the thinking process are probably more subtle, but I believe that they have plagued the efforts of logicians and mathematicians. And the errors of symbolizing are even more elusive, primarily because of the sheer difficulty of providing an adequate articulation of metaphoric knowing (e.g., the problem of what symbols "mean" and what qualifies as "universal"). Furthermore, none of these psychological processes operates independently of the others. That is, one does not conceptualize independently of sensory inputs and the process of symbol formation; nor do we perceive independently of concepts. In short, although the correspondences indicated in Figure 1 are oversimplified for purposes of analysis and exposition, they represent the fit between a given cognitive class and its parallel epistemological criterion.

Although psychologists have only begun the task of developing an adequate cognitive psychology, I must at least indicate what I am alluding to when I employ such terms as *conceptualizing, perceiving,* and *symbolizing.*

Conceptualizing: Cognitive processes that focus on concepts—their formation, elaboration, and functional significance to the organism; more deductive than inductive; focus is on the logical consequences of information currently available to the organism.

Perceiving: Cognitive processes that focus on observables—sensory inputs and their "meaning"; more inductive than deductive; focus is on the processing of sensory information.

Symbolizing: Cognitive processes that focus on the formation of symbols—"constructed productions" offered as representations of reality; analogical rather than deductive or inductive; focus is on the processing of

Figure 1. The basic paths to knowledge.

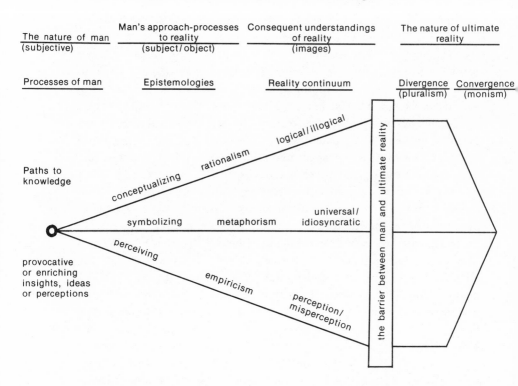

The nature of man (subjective)	Man's approach-processes to reality (subject/object)	Consequent understandings of reality (images)	The nature of ultimate reality
Processes of man	Epistemologies	Reality continuum	Divergence (pluralism) / Convergence (monism)

Figure 2. Representative special disciplines of knowledge.

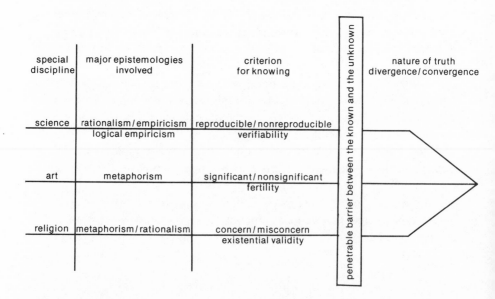

special discipline	major epistemologies involved	criterion for knowing	penetrable barrier between the known and the unknown	nature of truth divergence/convergence
science	rationalism/empiricism logical empiricism	reproducible/nonreproducible verifiability		
art	metaphorism	significant/nonsignificant fertility		
religion	metaphorism/rationalism	concern/misconcern existential validity		

"new-formation" (i.e., internally generated forms) rather than in-formation.

In Figure 1 I have attempted to show the relationships between man the knower and the nature of reality by means of three ways of knowing. The two columns to the right are separated from the other three columns by a barrier between man and ultimate reality. That which is epistemologically untestable lies to the right of this barrier and constitutes unknowable ultimate reality. That which is testable by some criterion for knowing lies to the left of the epistemological barrier and leads to "reality images" that are "true" or "real." Despite the efforts of great thinkers to somehow circumvent the epistemological limits involved, the only valid assessments open to finite man necessarily lie to the left of the barrier.

Such efforts to find truth have presumably been going on since man first made his appearance in the universe, and they have slowly evolved into the current special disciplines of knowledge such as history, literature, and biology. By definition, such specialties provide a highly selective view of reality and lead to divergent world views. The psychoepistemological basis for this state of affairs is depicted in Figure 2. For present purposes I suggest we ignore the right half of this figure and focus on the left side. It is to be understood that all three epistemologies are involved in each of the three representative disciplines of knowledge, but it is also clear that each discipline gives greater credence to one or more of them. The scientist, for example, "conceptualizes," "symbolizes," and "perceives" as a scientist, but he maximizes the rational and empirical ways of knowing and minimizes metaphoric symbolizing as a final judge. Conversely, the artist, who also invokes his entire cognitive repertoire, maximizes the symbolizing process at the expense of the conceptualizing and perceptual processes. There are, of course, wide variations in the possible combinations of epistemological profiles; this brief exposition should be taken as relative and typical rather than as absolute and general.

The essence of the theory of individuality, which takes its point of departure from factor theory,* is that a person's psychological makeup is an organized, multidimensional system. However, the total psychological

---

*Factor theory is a general scientific method for identifying the components of complex phenomena such as intelligence and personality. It involves an application of a modern algebra, the theory of matrices. See, for example, L. L. Thurstone, *Multiple Factor Analysis* (Chicago: Univ. of Chicago Press, 1947); and H. H. Harmon, *Modern Factor Analysis*, 2nd ed., (Chicago: Univ. of Chicago Press, 1967).

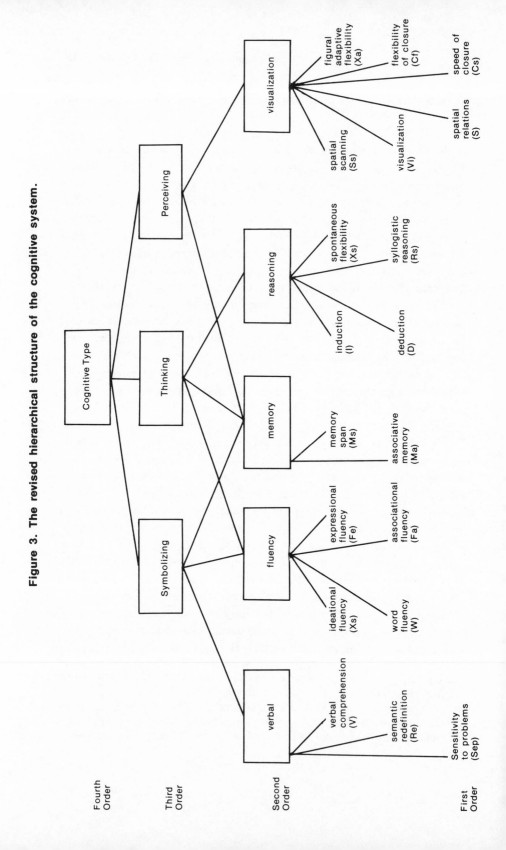

Figure 3. The revised hierarchical structure of the cognitive system.

system subsumes six subsystems, designated as sensory, motor, cognitive, affective, style, and evaluative. Highly oversimplified, the key idea is that individuality can be portrayed by subprofiles of subsets of dimensions for each of the six subsystems.

What are these dimensions, where did they come from, and how might they be organized? So far the theory includes some thirty cognitive dimensions and about twenty-five affective dimensions that appear to be hierarchically arranged. And they have emerged inductively from the hundreds of empirical reports in the factor-analytic literature. Let us look at cognitive structure as paradigmatic for each of the six subsystems (see Figure 3). What we have is close to two dozen first-order factors, five second-order factors, three third-order factors, and a fourth-order construct. The key concept is *cognitive type*, for it reflects individual profiles for the entire hierarchy of traits. Now, refer to Figure 4 for two simplified

Figure 4. Showing two persons, A (solid line) and B (dotted line), with the same IQ but with opposite mental ability profiles.

FACTOR	STANDARD SCORE
Number Space Reasoning Perception Memory Verbal comprehension Verbal fluency	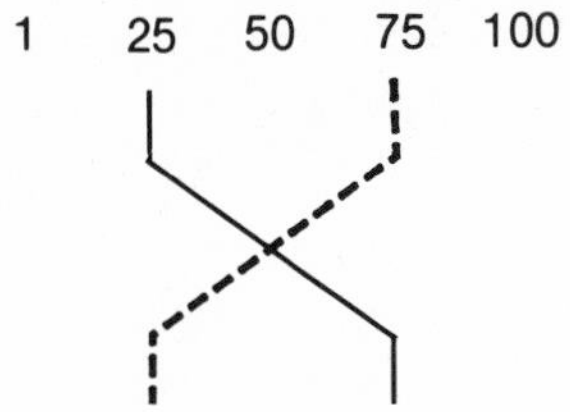 1 25 50 75 100

(i.e., showing only seven primary factors) cognitive profiles—A (solid line) and B (dotted line). The profiles show the relative strengths and weaknesses of two cognitive types, indicating which cognitive dimensions are optimally available to the subject in his interactions with the environment. Thus, the verbal type, A, will do better with verbal tasks and the quantitative type, B, will do better with quantitative tasks.

However, in this context we are more concerned with style structure and epistemic hierarchy. Thus, let us define the style system as *a multi-dimensional, organized system of processes (subsumes cognitive, affective, cognitive-affective, and epistemic styles) by means of which an organism manifests cognitive and/or affective phenomena.* Although the cumulative-data base in the style domain is such that it is not possible to present an empirically anchored picture at the present time, a hypothesized style structure that shows one logically possible hierarchical arrangement can be offered. This is summarized in Figure 5.

Style is *a characteristic mode or way of manifesting cognitive and/or affective phenomena,* and cognitive style is *a style construct that is limited primarily to cognitive processes.* The construct of epistemic style is defined as *a style construct that simultaneously invokes a valid truth criterion (i.e., leads to a justifiable knowledge claim in addition to being a characteristic mode or way of interacting with the environment).* The important concept here is epistemic type, which is a profile or a hierarchy of all style con-structs but especially of the three epistemic styles indicated at the second level.

If we now embed both the cognitive and the style system within the total psychological system, the individuality basis for world view should become more apparent (see Figure 6). Our focal point is on world view at level five. According to the hierarchical structure on the left side of this fig-ure, differences in world view can be accounted for in terms of variations in cognitive and epistemic type. That is, the rationalist (in varying degrees of purity, of course) will manifest a typical style profile, as will the empiri-cist and the metaphoric. And, because there are also (unknown) relation-ships between styles and cognitive abilities, it is to be expected that dif-ferent subsets of these dimensions will also be part of the total psycho-logical profile (e.g., affective and value dimensions as well) of the meta-phoric, rationalist, and empiricist. This epistemic hierarchy, combined with cognitive profile, is the major determinant of world views.*

---

*Values and affect also determine world view, but, according to the position here espoused, not as directly. Their influences filter down from the top (through cog-nitive-affection integration), although there probably are direct value influences as well. In short, all six subsystems (the sensory and motor systems were omitted from Figure 6 for the sake of simplicity) are in constant interaction; however, it is our guess that values and affect are of greater importance for the understanding of ideology, whereas cognition and styles are directly involved in the development of world view.

Figure 5. Hypothesized epistemic style structure.

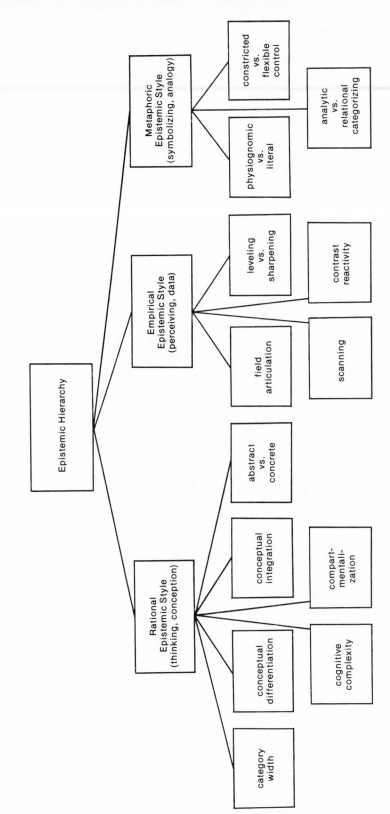

The specific linkages of each epistemic style to cognitive abilities and affective traits remains to be spelled out (see Figure 7 for examples). However, it is anticipated that epistemic styles will be primarily cognitive in nature (i.e., with few affective linkages).

THE EFFECT OF DIMENSIONAL AND ORGANIZATIONAL SHIFTS

For reasons not clear to me, when we speculate about extraterrestrial life we tend to attribute to those possible life-forms a degree of intelligence higher than our own. It would be highly unlikely, however, for this to occur for *all* the components of intelligence. What is more likely is that different types of beings would be characterized by different profiles of strengths and weaknesses.* It is, of course, conceivable that the hereditary and environmental determinants occurring in a particular galactic locus could interact in such a manner as to produce a super-race—a subgroup of beings who perform at a high level on *all* the cognitive components. However, it seems more probable that there would be a complete range of individual differences on each of the cognitive components.

I can see three major subclasses of dimensional and organizational sources of intellectual variation.

1. *Variation in any one dimension or in a specifiable subset of dimensions.* This is the most important source of individual and group differences in intelligence, and it lies at the core of the multifactor theory of intelligence. It says that each person's intelligence can be characterized as a multidimensional profile (see Figure 4). The idiot savant is the prototype of super development of a single cognitive dimension. He is able to accurately recall long serial lists of digits. Memory span is the most likely factor counterpart of this ability. However, the idiot savant's intelligence is typically limited to superdevelopment of a small subset of intellectual components. The more typical intelligence profile involves strength (and weakness) in larger subsets of cognitive dimensions. For example, the cognitive hierarchy depicted in Figure 3 indicates that high-scoring "perceivers" will score well on the lower-order memory and visualization factors but poorer on the remaining dimensions. "Conceptualizers," on the other hand, should do relatively well on the fluency and reasoning dimensions but not as well on the visualizing components.

2. *The emergence of new dimensions.* New dimensions come into play as a result of organismic change over time. Organismic change occurs both phylogenetically and ontogenetically, and the implication is that new dimensions are the structural basis of increases in organismic complexity.

---

*In a manner comparable to the verbal and quantitative types shown in Figure 4. However, because of the possibility that life on other planets could have evolved quite differently, the types may not exist on planet earth.

Figure 6. The hierarchical structure of personality.

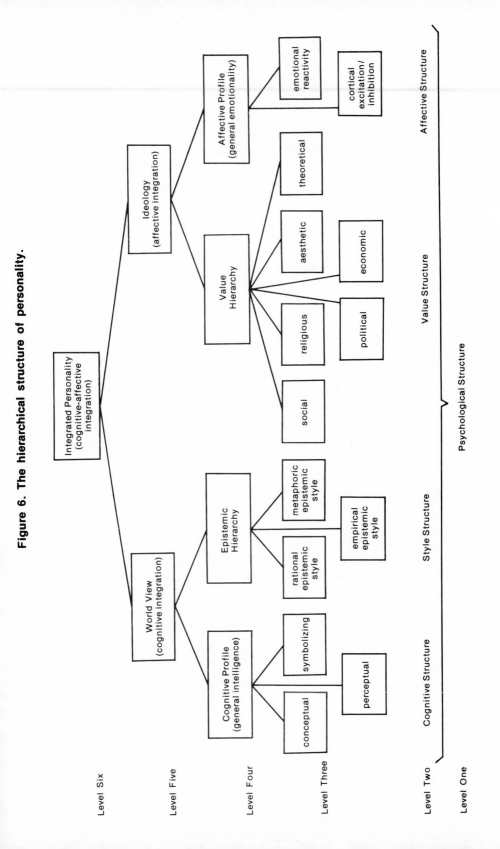

For example, a structural description of the human mind will require dimensions not present in a dog's mind, and similarly, a structural description of a dog's mind will require dimensions not present in the mind of a rat. Since language is unique to humans,* the various linguistic components of intelligence (see the verbal and fluency factors in Figure 3) constitute the most obvious case in point.

It is, of course, quite possible that extraterrestrial beings have evolved beyond man in certain intellectual components. Take, for example, the sensory components necessary for seeing such light wavelengths as the cosmic, gamma, ultraviolet, infrared, and Roentgen rays. Since the vision of Homo sapiens is limited to one-seventieth of the light spectrum, it seems reasonable that the vision of other hominids might be better. Furthermore, while it is true that man surpasses lower animals in a wide variety of ways, lower animals also surpass man in many ways. Obvious examples include sense of smell, locomotor speed, strength, sound discriminations, the swimming superiority of fish, and the flying superiority of birds. Extrapolations to future terrestrial evolutionary developments or to extraterrestrial evolution are legion. Thus, organismic tele (i.e., distance) vision is a plausible notion, as are the possibilities of cognizing an n-space world (assuming it exists somewhere) instead of a three-space world, cognizing time intervals accurately, the varieties of extrasensory perception, and being highly (accurately) sensitive (superempathic) to the feelings of others.

Furthermore, evolutionary development could have taken an entirely different course in other cosmic environments. For example, what might a manlike creature be like if he evolved in the context of radial rather than bilateral symmetry? It is conceivable that radial symmetry would have involved an even more efficient central nervous system than we have. For example, bilateral symmetry involves two interacting brains (the left and right hemispheres); a radially symmetric brain would be anatomically unitary and ideally suited for coordinating inputs from any direction. Since a radially organized man would not have a front or a back, he would be able to accommodate stimulus inputs from all directions. If such an

---

*While it is true that the recent experiments on language acquisition in chimps (e.g., Washoe, Lana) raise a question about "homo symbolicus," the point being made here is not challenged thereby for the simple reason that chimp linguistic competence to date is, in fact, paltry when compared to that of man. Even more crucial, there is no known linguistic-cultural evolution comparable to that of man, and that is really the point.

organism were moving in all the possible directions of three-space, rather than in our typical two-space (i.e., the plane of the ground), a radial system of sensory intake would be necessary. Radial motor outputs (e.g., a multitude of limbs, perhaps roughly like legs with the prehensile characteristics of the hand) would also be required. Having such characteristics would probably constitute at least the beginnings of cognizing an n-space world. Furthermore, everyday movement in a three-space world would not represent a very great departure from our typical two-space movements. That is, such an extrapolation is minimal and not improbable if there is life elsewhere in the cosmos.

Another interesting dimensional variation would involve memory components without a built-in forgetting component. One's first reaction is that this would be an enormous adaptive gain, for the simple reason that all experiences would be cumulated and recallable for future use. Thus, the potential knowledge available, especially for those with formal education, would be enormous. That could be desirable in cases where the cumulated experiences are either neutral or positive. But what about negative experiences? They would also be retained in this super memory. Thus, a traumatic experience could easily result in a continuing trauma, or at least a reliving of the initial event. Furthermore, one wonders about the possibility of overload in a memory system that does not forget. This suggests the seemingly unlikely conclusion that an imperfect memory mechanism (such as that of man) is more adaptive than a perfect one.

3. *Variations is how the dimensions are organized.* Another source of variation in the performance of intelligent beings is in the organization of the cognitive components. A relatively simple organizational scheme, J. P. Guilford's cubical model, postulates that each of the cognitive components functions totally independently (i.e., orthogonal) of the others.[5] However, it is doubtful that this model can deal adequately with the more complex aspects of cognition. The hierarchical model, which implies correlated dimensions and higher-order integrative factors, seems more adequate for this situation. The point is that the higher-order factors can differentially channel the necessary information processing to lower-order cognitive processors. Thus, the hierarchical mode of organization provides for more sophisticated coordination as well as specialization of function. Information-transmission mechanisms, such as feedback loops and simultaneous processing, are more flexible and efficient in providing an adequate account of how cognitive components might be organized in the most intelligent life-forms.

EPISTEMOLOGICAL IMPLICATIONS

The evolution of new cognitive dimensions and/or patterns of organization has epistemological implications. For new modes of cognizing have the potential of opening up new vistas of reality, thereby providing us with new and/or enlarged world views. Intuition and empathy are cases in point because they represent gray areas in the realm of human knowing—that is, there is convincing evidence that humans can intuit and empathize, but our knowledge of these processes is so limited that we do not trust the validity of intuitive and empathic awareness. The opportunity to study these processes in highly developed subjects (extraterrestrial subjects?) would provide some of the answers. Thus, we return to the point at which this essay began—the issue of man's consciousness and how it might be expanded.

The existence of extraterrestrial intelligence would have an enormous de-encapsulating impact on man's world view, perhaps of a magnitude comparable to the Renaissance and the subsequent Copernican revolution. However, if it turns out that intelligent life is limited to earth and that the contributors to this book have merely contemplated the possibility of ETI, we still stand to gain in our view of the way things are. For if man wishes to develop a more inclusive world view he will have to break through the several cocoons within which he is encapsulated. Of course the first step in this process is to recognize the ways in which he is encapsulated. The theme of this essay is that the source of encapsulation, or limited consciousness, is in our depreciation of those modes of knowing that are alien to our personal epistemic commitments. Thus, the super-rationalist tends to overestimate the importance of thought and to underestimate the observables of the empiricist and the insights of metaphorism. Similarly, the superempiricist wants facts only—at the expense of the reality-expanding analogies of the metaphorist and the ideas of the rationalist. And the supermetaphorist responds to the symbolic-metaphoric universes of the arts and the humanities at the expense of the factual world of the empiricist and the logic of the rationalist.

The antidote to the limitations of epistemic specialism is obvious: encompass *all* of the ways of knowing to the fullest extent possible. This involves not only developing the ways of knowing that are low in one's epistemic hierarchy, but also further developing primary epistemic commitments. Thus, it is a question of making up for both a qualitative (the lowest epistemology in one's hierarchy) and a quantitative (upgrading performance in each epistemology to the limit of one's capacity) deficiency. This is, of course, a tall order. The price is hard work; the reward is expanded consciousness.

Let us assume that there are individuals willing to make the necessary effort and that they achieve levels of performance close to their limits on each of the ways of knowing. Let us further assume that these individuals are capable in all of the ways of knowing—that is, that their overall level of intelligence is high and that, with effort and training, they are capable of high performance levels on each of the required cognitive abilities. Although each such person would still be encapsulated in the sense that his performance would be less than perfect, it seems clear that he would be relatively unencapsulated.

Thus it follows that these individuals should be capable of generating relatively penetrating and broadly inclusive world views. In short, they would be optimizing their consciousness. I have characterized such unencapsulated persons as persons of individuated consciousness (see Figure 7).

Figure 7. The awareness continuum.

Omniscience ⎯⎯⎯⎯⎯⎯ Ultimate consciousness (?)
Unencapsulated man ⎯⎯⎯ Individuated consciousness
Encapsulated man ⎯⎯⎯⎯ Specialized consciousness
Mass man ⎯⎯⎯⎯⎯⎯⎯ Collective consciousness
Sentient structures ⎯⎯⎯ Threshold of consciousness
Symbolic forms ⎯⎯⎯⎯⎯ Personal unconsciousness
Primordial images ⎯⎯⎯⎯ Collective unconsciousness
Free energy ⎯⎯⎯⎯⎯⎯ Ultimate unconsciousness (?)

The implication of Figure 7 is that there are varying levels of consciousness, ranging from the level of unorganized or entropic energy (i.e., the winding down or using up of energy, which is a consequence of the second law of thermodynamics)—labelled ultimate unconsciousness—to the ultimate in consciousness—which is given the suprahuman label *omniscience.* The implication is that various life-forms have evolved out of the inorganic world, and further, that the level of consciousness varies with the phylogenetic level.

Although man is at the apex of the animal kingdom, at least as we know it on planet earth, Figure 7 indicates that there is considerable varia-

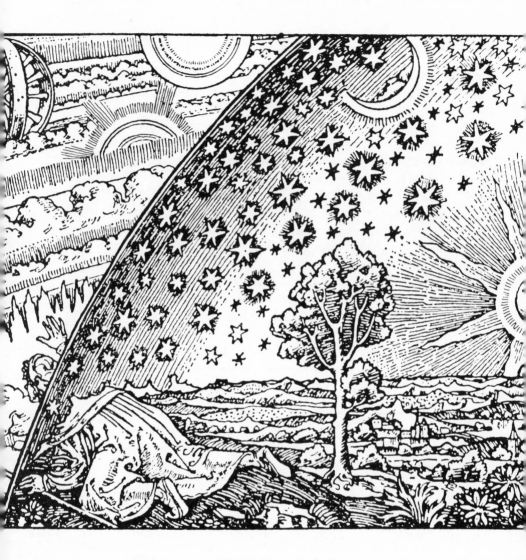

tion in the level of consciousness that individuals achieve. Thus mass man, or sheeplike man, merely follows the crowd. He is satisfied with a collective level of consciousness. Encapsulated man, although very limited in absolute terms, is, nevertheless, a man of great awareness—the specialized awareness of his major epistemological commitment (i.e., rationalism, empiricism, or metaphorism). It is only the relatively unencapsulated man who achieves the level of awareness labelled *individuated*—meaning that level of consciousness which sees maximally by means of all the available ways of knowing. By individuated consciousness I mean that degree of consciousness which runs the risk of coming into contact with too much reality. Such consciousness has been variously described by experts in sociology, psychology, religion, and philosophy. The philosophers and religionists have tended to view this level of awareness as a state of enlightenment. In the West we think of such men as "wise"; in the East such states have been referred to as *satori* (Zen), *samadhi* (Hindu), or *nirvana* (Yoga). Sociologists and psychologists have tended to see such states as characteristic of such outsiders as the alienated and the psychopathological. I suspect there is an element of truth to each of these views. The multi-epistemic world viewer does run the risk of "seeing through" in the manner of the outsider (i.e., becoming alienated), thereby exposing himself to the risks of seeing too much. However, my argument is that this kind of risk is worth taking because of the greater awareness of individuated consciousness.

Are there levels of consciousness that go beyond individuated consciousness? Eastern mystics would probably say yes. And, while I am skeptical, we in the West ought at least to listen. And what about ETI? My guess is that, if there is a high level of intelligence elsewhere in the cosmos, it also would be short of omniscient. It may be individuated at a similarly high level, but the epistemic routes to awareness would probably be different. And it is that possibility which undoubtedly intrigues us so much about ETI. For if ETIs have different psychobiological mechanisms for knowing, then the results of an interaction between ETI individuated consciousness and planet earth individuated consciousness could indeed result in a superindividuated (even though subomniscient) consciousness.

Let us imagine such a man—cosmic man, a product of the individuated consciousness of at least two galactic supercultures. He would not be a man of the world, he would be a man of the universe. His first loyalties would be to the central government of the universe, followed by the decentralized governments of the various galaxies and planets, and then, possibly, to such intraplanetary regional governments as the nations of the world. Because of the experience of intergalactic and interplanetary space travel, cosmic man would have had direct experience with the principle of relativity, not only in terms of time and space, but also in terms of being. His outer-space view of earth, for example, would be roughly comparable to the earthman's aerial view of a major landmark in a city—a mere point on the vast expanse of the earth's surface. And if even a segment of the cosmos is capable of sustaining living forms, it is highly probable that the cosmic range of species variation would be greater than that of a single planet. And cosmic man would have been in contact with this variety of life alternatives.

But what about the problems of communication across cosmic cultures (assuming they exist elsewhere)? This could well be an impasse, for to the extent that such cultures differ from each other (as a result of the different environments to which evolving life adapted), communication may not be possible. If languages, for example, are very different, how could the gap be bridged? Let's extrapolate from similar problems on earth. How well do we understand the signals of other animals? There are intelligent life-forms on planet earth—subhuman primates and dolphins, for example. After all, many subhuman mammals have well-developed brains, and adaptive capacity is a function of brain development. How deeply does the psychologist get inside the mind of other animal species? How well do we communicate with these species? Although the examples from lower animals are primarily nonlinguistic, they exemplify the kind of problem that would probably exist across cosmic cultures. A major reason for taking this stance is the high probability that extraterrestrial intelligence is more likely to be manifest in a species other than earthlike Homo sapiens.

Let us push this bit of fantasy still further by assuming an extraterrestrial species with greater intelligence than that of man. We would be dependent upon their superintelligence for figuring out how to communicate across cosmic cultures. If such communication occurred, man's major adjustment would be the relatively minor one of learning how to accept a secondary role in the universe. However, if communication between man and a superior cosmic culture turned out to be impossible, man's problem would probably look more like that of some near-extinct species of animal, because the cosmic superbeings could decide, in a manner paralleling man's behavior on earth, to preserve a small population of Homo sapiens on an earthly game reservation. The idea would be to keep a breeding population going so that specimens from this relatively rare cosmic species would not disappear. Of course, a naturalistic setting would be provided in order to permit the full range of human expression in the hope of eventually breaking through to an understanding of such a fascinating and surprisingly highly intelligent species. The term *surprising* would have to be used because of the enormous number of human contradictions and paradoxes. For example, cosmic scientists might observe that humans make enormous sacrifices to build schools and universities, but that most of them end their education at about the age of sixteen. And cosmic man might well be confused by the human practice, done in the name of democracy, of spending millions of dollars to elect politicians who have no special training or qualifications for office. Finally, there is the human practice of sparing no expense in trying to save or prolong the life of one sick or old person, and, at the same time, sparing no expense in the conduct of wars, which involve the slaughter of millions. Of course the inventory of human foibles is much longer. And man can also point to his achievements. However, man's most outstanding accomplishments are technological, and, in spite of the impressiveness of those advances, cosmic man might well doubt the overall intelligence of life-forms on planet earth, judged by the record of its most intelligent species.

NOTES

1. See J. R. Royce, *The Encapsulated Man* (Princeton, N.J.: Van Nostrand, 1964).

2. J. R. Royce, "Cognition and Knowledge: Psychological Epistemology." In E. C. Carterette and M. P. Friedman, eds., *Handbook of Perception,* Vol. 1 (New York: Academic Press, 1974), pp. 149-176.

3. J. R. Royce, "The conceptual framework for a multi-factor theory of individual differences." In J. R. Royce, ed., *Multivariate Analysis and Psychological Theory* (London: Academic Press, 1973), pp. 305-407.

4. J. R. Royce, "Cognition and Knowledge"; and J. R. Royce and Herman Tennessen, *Inquiries into a Psychological Theory of Knowledge* (in preparation).

5. J. P. Guilford, *The Nature of Human Intelligence* (New York: McGraw-Hill, 1967).

"It is becoming increasingly clear, in this second half of the twentieth century, that moral sentiments are radically changing, the sphere of consciousness is enlarging, boundaries are dissolving, and our earlier certainties—even the certainty that certainties are possible—have been swept away. To find some sort of foothold on the shifting terrain and in face of the changing landscape, *we need a new definition of man.*"

So writes Wilfrid Desan in the introduction to the second volume of *The Planetary Man* (1974). In that and his other works Desan is a philosophic futurist who attempts to see and assess man realistically in an ever larger society. In *ETI* he applies this concern to "biocosmic man."

Dr. Desan is Belgian by birth, French by education, American by adoption, and Planetary Man by choice. Among his writings are *The Marxism of Jean-Paul Sartre* and *Tragic Finale: An Essay on the Philosophy of Jean-Paul Sartre.* He is a Harvard alumnus and at present teaches philosophy at Georgetown University.

"The notion of a single truth has been useful to science, but when fully observed—that is, including an extraterrestrial range—the notion loses its firm univocal applicability. Knowledge on Planet X as a whole or on earth as a whole is only angular. *"*

Wilfrid Desan

Angular Truth and Planet X

Recently I reread Aquinas' treatise on the angels, some one hundred pages of perspicacious speculation.[1] It shows a sharp insight on the part of Aquinas into the way of knowing among men who are caught in matter, through the hypothesis, albeit unreal, of beings that are not made up of matter. In this exercise the unreal clarifies the comprehension of the real. I could not help but think that if the medieval theologian could write so abundantly and confidently on what he had neither seen nor heard, so we too, earthbound as we are, might write about another type of stranger, less spiritual perhaps, but more in line with modern speculation—namely, the beings who live on another planet. We might even dare to speculate upon how they know, what they know, and how their truth should be understood.

Of course I do not know with certainty whether or not there exists a "Planet X" where people live and know; neither did Aquinas know whether angels exist. The content of this article, although plausible in form and content and as realistic as I can make it, remains hypothetical. Yet in writing it I discovered what Aquinas had seen long ago, that the consideration of the hypothetical is useful, and in some strange way it illuminates the real. For it may well appear that speculation upon the possibility of knowledge and truth thousands of light-years from here will serve to undercut the pride and self-sufficiency of terrestrial man; we may come to see that what he knows is minimal and the way in which he knows it is very one-sided.

THE GREEK UNIVERSAL AND ANGULAR TRUTH

We cannot philosophize on knowledge without taking into account the structure of life itself, here and on Planet X. Yet if it seems logical to start with the structure of life—as did Aristotle when he discussed the problem of knowledge in his book *On the Soul*—the reverse is equally true: we cannot elaborate on the nature of life without making certain that we can handle the problem of knowledge. For the danger always exists that in the very act of attempting to know Planet X and the form and nature of life on that planet, we are *distorting the reality that we want to know*. Our observation itself may be altering the things that are observed in the very act of attempting to know them.

At the core of our attempt to understand knowledge must be a certain readiness to see it as something that has great flexibility. When we say that extraterrestrial beings are of such and such a nature, or know in such and such a way, we should not claim that this nature or this knowledge are the same as ours. A prerequisite to the understanding of other planetarians is a break with the rigidity that, it seems to me, has always dominated our own way of knowing.

For centuries the philosophical study of the act of knowing has been guided by the work of Greek thinkers, in particular that of Plato and Aristotle. What characterizes their approach and still dominates the scene is what I would like to call the *reductio ad unum*. This is tantamount to saying that the only way of knowing, or of real Knowing, is to discover the ultimate Law. What matters is to explain the many by the *One*. The wise man is the sort of person who, after long observation, or perhaps by sudden intuition, has discovered exactly that: he knows more than one particular event in human behavior, more than one case; he knows them all because he has the key to the enigma that dictates human behavior.

The shape and form of that *One* may vary. The religious man finds God the ultimate and absolute One. The scientist finds the ultimate law in physics and mathematics. What matters in both cases is the acceptance of a single Entity or a single Law that explains the multiple visible around us and makes it intelligible. This is considered to be knowledge.

Both Plato and Aristotle say things in this way. Plato's Idea is a *definition* of the many: it reveals what the many have in common. Gertrude Stein's answer to the question, What is a rose?—a rose is a rose is a rose—would have made no sense to the Greeks. On the contrary, to them to know what the many (roses) are is to reduce them to the one, through the revelation of the very essence of roses, or that by which all roses are roses. Aristotle's "universal" is itself the idea as communicable to the many roses to which it can and will be applied. What strikes us here is that in some way *the One is present in the diverse*. This is of major significance. The One in this case has a character of *sameness*—the universal is the same for all—and when we talk about a rose we all mean, or should mean, the same thing, with the result (at times the unfortunate result!) that the one-and-the-many becomes the one-is-the-same-as-the-many.

The universal and incessant attempt at clarification resulting from the Greek method has paved the way for the development of science in the Western world. Science is a field dominated by the *sameness* and the *one*. Mathematics by its very nature has a sameness that eliminates all diversity. In the grasp of a mathematical equation lies a universality of understanding, independent of the individual perceiver: mathematics in Washington and Peking is the same.

Clearly this is not the case for all forms of knowing. Take, for example, an abstract term like *justice*. In most cases the understanding of that word is at best partial, and more often than not, highly personal. The individual understands it in his own way. Yet we know what justice is, even though few among us would be able to define it in such a way that all would agree with the definition. This does not prevent everyday life from running smoothly. However, this *personalized* understanding of the meaning of a term does in fact throw some light on the way individuals *know*. Individuals know *diversely*, yet they still *know*. This is taking us away from the Greeks, with their clarity and precision, and onto new terrain.

Having spent World War II as a movie cameraman, I have a clear memory of the constant worry of the cameraman, which is how to contain the set, and nothing but the set, within the viewer. One must choose the right lens for the right task—the wider the angle, the larger the view; the longer the lens, the more narrow the field it covers. When the set or the person is photographed through a wide-angle lens instead of through a telelens it not only produces an alteration in the amount of space the lens covers but it also brings about a qualitative difference, for the shape of the face is not the same in the final result. All photographic vision is angular, in the sense that it is caught within the angle of a certain optic. It is this term that at present I transfer to the vision of the individual man or woman approaching reality. Their vision, too, is limited and qualified: it is *angular*.

In using the phrase *angular vision*, with the individualization it implies, I am moving away from the Greek concept of the *universal*, with its heavy emphasis upon the similar. Perhaps it is the familiarity with a geographical diversity as well as with a much longer history (a perspective missing to the Greeks) that has compelled us to devise a method that abandons stress upon the common in favor of stress upon the diverse and the unequal. Clearly, within that variegated ambience that both geography and history reveal, the vision of one man contains certain elements that the vision of another lacks. A person is made an individual by the triple dimension of past, present, and future. Genetics has contributed to mold him into what he is. The ambience in which at present he moves and the collective of which he is a part drastically affect his physique and his mental makeup; yet it does so diversely, and makes the one so and the other such, both in being and in thinking. Perhaps more than anything else it is the future that individualizes a man. In the light of that future, which in most cases is an individual concern, ideas are formed and re-formed; for the extent to which the comprehension of a particular idea affects my own survival alters the comprehension itself. Each individual, moving at the crossroad of so many influences, uniquely so, *thinks* uniquely and differently. My ideas are constantly remolded. Yet, notwithstanding these fluctuations, or perhaps because of them, I often get hold of the truth. There is a conformity with the real—what has traditionally been called truth—but this truth will be an *angular* truth.

We may conclude that this truth will be both unique and correct: it belongs to me and to me alone. It does not follow that this truth cannot be shared to an extent, nor does it follow that my truth is complete, exhaustive, and all-encompassing. Nor does it mean that, having my share of the truth, I might not be wrong in many other ways.

THE GROUP AND ANGULAR TRUTH

Angular truth is the vision of an individual. In keeping with the same semantics, I shall call the vision of the group—in other words, the truth of the collective—a *generic truth*.[2] Just as there is knowledge or a way of looking that belongs to an individual, likewise there is knowledge that belongs to the group. Clearly a certain group has its own way of looking at things, and, as an unavoidable result, it discovers a truth that is not the way another collective or group looks at reality. A Chinese vision of the world is not the same as an American way of looking at things. The knowledge of the Chinese is unique because their very being-what-they-are is unique. I shall not claim, of course, that cultural groups—nations, or any other forms of the collective—have nothing in common. I merely want to state that their apprehension of reality is diversified; and yet, notwithstanding its diverse nature, it may well be a truthful apprehension of reality.

The vision of a group, of course, fluctuates; within its midst an incessant correction is at work. Yet because of this corrective function within, the *generic* truth opens a vista that far transcends the vision of the individual, but which sadly enough escapes him. The many failures of individual men and women result from ignorance of what the Totality knows.

Collectives are numerous upon this planet, be they called races, or tribes, or nations, or groups. To define them is not our purpose. What matters is the insight that they are in many ways different in their being and in the quota of the truth that they hold. If we were permitted to add together the many *generic truths* of the many collectives on this planet earth (considering the overlap, that is, not an addition in the strict sense of the word, of course), we would obtain a sum total of knowledge and truth that in turn could be called a generic truth of the human totality as a whole.

However, in this hypothesis it is natural that the different collectives making up the human totality become by the same token *angular* in their vision when compared to the larger volume of the real as seen and comprehended by the human totality as a whole. But even the vision of the human totality, however impressive and overpowering when its achievements are added together, would in turn fall to the level of the angular if and when a truth and a vision of the universe would emerge, representing a knowledge far beyond the knowledge of the terrestrial totality by itself. A natural conclusion would be that the truth possessed by an extraterrestrial collective, such as the one living on Planet X, would within the universal acquisition of truth become the carrier of what we have called, with good reason I believe, an angular truth. The inhabitants of Planet X, too, have *their* truth, which most probably is fragmented, just as ours is fragmented among the many components of a human totality here on earth.

We can see now how the concept of angular truth becomes precious to us in the study of extraterrestrial life and knowledge. We are now better equipped to continue our research, for we shall keep in mind what we can and cannot do. The knowledge of the extraterrestrial people may fall partly within a universalizable dimension, like the one we have inherited from the Greeks, but it may also fall outside this self-imposed discipline. The need to discover that unknown element compels us to proceed with care.

Unfortunately we have at our disposal only a semantics that is forged out of and remains *our* experience, and only with great hesitation can we track down the modes of knowledge of entities that lie in the penumbra of immense distances from here and that may or may not feel and know, hope and wish, love and die the way we do. This is where the *angular* vision comes in, for it contains the flexibility we need and consequently the potential of englobing the mystery of their noesis.

CONCERNING THE NATURE OF LIFE ON PLANET X

We can only speculate about knowledge on Planet X if we speculate on the general way its inhabitants live and relate. From the start we have a full right to posit that Planet X, with everything that is on it, being part of the universe, is caught in the dimensions of space and time, like the rest of that same universe. The notion of time is perhaps more complex than that of space, since the former admits of more possibilities of variation. Yet, whatever form of time exists in that strange land, we would have, in combination with space, a quadrant wherein things of matter, as perceived by the inhabitants of Planet X, are measurable, just as they are on earth. This of course would not be the case if they (or we) were made out of nonmatter, or what is sometimes called spirit. Without denying the possibility of the latter, let us limit this discussion to the observation of life as we deem it possible among the quantifiable and the organic. The sole postulate of this essay is that there is a form of life comparable to or better than ours but not yet pure spirit. This species could be either unfragmented or fragmented; if fragmented, it could be so either in space only or in space and time.

A first hypothesis that cannot a priori be excluded is the existence of an entity that is alive yet unfragmented. It is constructed by way of symbiosis, composite and unified into a single entity. Its receptive powers, such as sense perception and perhaps even intellectual performance, are not distributed among separate beings the way they are on earth, where the totum is broken up into separate components called individuals. Birth and growth take place within the extraterrestrial totality in a cellular way, and death is nothing but the evacuation of cells, such as we observe in plants. Logically that strange being would exclude anyone else within its own species: it would be a species unto itself. As a result the problem of communication would be nonexistent, yet a sense apparatus would be present, geared toward the absorption of the inorganic. It would be fitting to endow that massive organism with a consciousness. One consciousness would suffice. Several intellects could be there, but all of them would be integrated by a central awareness, somewhat in the way neurons in the brain are united by an individual consciousness. The existence of such a huge monad (with windows!) is of course speculative; yet it cannot a priori be excluded from the realm of the possible.

A fragmented totum seems more probable to us. In that case there would be more than one component within the totality. The advantages are obvious. First, there would be a much greater resilience to destruction, since the death of the one would not be the death of all. Second, a multiplicity of components within one totum would allow better adaptation to the multiple resources that I shall presume to be on Planet X. I shall return to this point later.

If the fragmentation is one of space only, without fragmentation in time, then we have a juxtaposition of living beings, each of them going on without interruption. There would be no need for reproduction, since replacement would be superfluous. These organic realities would be asexual and endowed with some form of immortality. They would move independently on a spatial level, yet for their subsistence would be interdependent. This hypothesis cannot be considered impossible; yet it appears less probable than that of a totum fragmented both in time and in space.

This is, of course, what we see on earth. If this is also the case on Planet X (my next supposition), then survival takes place through replacement, and there is no more place for immortality than there is on earth. There will be destruction through death and the promise of new life through birth. Although procreation without sex cannot be considered impossible, the inhabitants of Planet X are most probably sexual. If they are, it is their privilege to love and to die. The ancients saw this long ago and rightly concluded that the gods do not love.

A structure fragmented in time and space naturally excludes the physical compactness that was an essential feature of the first hypothesis, but it does not cancel oneness of origin. This is an important point to keep in mind in an era when taxonomists have stressed the similar and the common in their classifications. The classification of organisms into genera, species, classes, and so on, results in building up a form of oneness of a logical nature; yet this should not make us lose sight of the fact that the ultimate ground of commonality is their genealogical oneness in origin. There and there alone lies the intelligible explanation in depth of the one-and-the-many. The fundamental oneness of the *many* (organisms) lies in the *one* beginning, not in the universal. The very transmission of life from the beginning is present in the "descendants" and ensures the *identity* of life. Within that identity *diversity* of life is protected as well, thanks to bi-sexual genetics and the resulting mutation of the genes.* Change, there-fore, is a capability of life on Planet X.

Turning now from the concept of time to that of space, we note that upon landing on Planet X, we shall expect to confront a multitude of inter-related living things. This may appear to be a rather obvious remark by now; yet it is an important one. The notion of a "multitude" implies the existence of a collective spread open in space. Their interrelatedness is not merely a biological common origin but also a way of being, for although they are separate and distinct, they need one another. Planetarian X, born from Planetarian X, leans upon its fellow beings in order to be. Depend-ence in origin transforms itself into dependence in subsistence. Space is the ground for the ecological milieu and niche wherein this activity flourishes.

This ecological ambience has molded a particular collective or totum on Planet X just as it has done here. The Planetarians X, although consti-tuting a global totum on their planet, may very well be broken up into smaller yet sizable subtotums. The original "oneness" perdures, no doubt; yet survival promotes alliances within the global totality, and the emerging

---

*We cannot a priori exclude the possibility of change within one individual without sexual interference. But the acceptance of a bisexual genetics makes this hypothesis superfluous.

of smaller groups is a natural phenomenon.

What is important is that the Planetarian X totality be seen as a coherent entity. Hence when we compare Planet X and earth, we must not compare and oppose individual John with his counterpart on Planet X, individual Nhoj, all by themselves. We can, of course, do so, but we must keep in mind that they are only such or so because they are part of two immensely different totalities. The inhabitants of Planet X have reacted as a group upon an ecology that is very different from ours, and in reacting they have chosen, perhaps with a certain amount of freedom, to be what they are. Whoever belongs to a totum is radically shaped by it. Hence if John is different from Nhoj, it is because John belongs to a collective that is different from that of Nhoj.

At the same time the diversity must and will be protected even within the same collective. The very fact that we give Nhoj a name must remind us that he is an individual: he is part and fragment of the totality, yet he is so uniquely. He is himself. We can now see how the fragmentation that we have accepted on a space-and-time level in our third and most probable hypothesis introduces more diversity within the totum—whether Planetarian X or terrestrian—and, by the same token, introduces greater biological adaptability and resilience than a type of unfragmented totum would ever be able to exhibit. For it is most likely that on Planet X a variety of elements will be present that are both the source of life millennia ago and its means of survival. On an expanse replete with diverse resources a diversity of inhabitants will adapt itself better than a homogeneous bloc. The fragmentation of the totality provides this internal diversity.

CONCERNING THE KNOWLEDGE
OF THE INHABITANTS OF PLANET X

There is a beautiful Augustinian distinction taken over by Aquinas in the *Summa*, that between *cognitio vespertina* and *cognitio matutina*.[3] The *cognitio matutina,* or "knowledge of the morning," is allotted to the angels, because in a world of nonmatter or of the nonquantifiable they do not feed their knowledge through the senses, since they have no senses. Their knowledge is *infused* and comes to them directly from God. The *cognitio vespertina,* or "knowledge of the night," is that which belongs to men. This knowledge is obscure and lacks clarity because it is obtained through the senses; Augustine, with his distrust of the senses, once more gives the *civitas terrestris* a raw deal.

The Planetarians X are not angels or spirits. We do not know with great precision what we are saying when we talk about angels, but let us exclude them from Planet X, where the inhabitants live on a spatial body and learn *through* a medium called body, with its sense organs, *from* a world of quantifiable matter. The possibility of some innate knowledge is neither included nor excluded. Even if we do accept the innate awareness of some fundamental principles, such as, for example, the principle of contradiction, this still would not affect the ground rule that says that in a world of matter one knows through matter, although *that which knows* might be more than matter or, if matter, at least matter in a different form from ordinary matter.

In allotting the *cognitio vespertina* to the Planetarians X—Augustine is no doubt turning over in his grave upon hearing this surprising transposition from the *Civitas terrena* to Planet X—we must endow them with some form of sense perception. Since, by hypothesis, the Planetarians X have reached a stage of evolution comparable to ours, they must have a fairly refined sense apparatus. The Aristotelian claim that there are five senses and five senses only and that through them we are aware of everything external is debatable, even on earth, according to modern science. It would be presumptuous, in any case, to apply these observations without qualification to Planet X, which has never been the object of our empirical survey. But looking at the situation philosophically one sees that, since we must consider survival as an essential aim, it is fair to inquire what is necessary to survive as a higher species.

Strictly speaking, no one sense modality is an absolute requisite; yet it looks as if some sense of touch is necessary in a world made up of organisms moving within a limited space. Touch constitutes a permanent warning system. Trees do not feel, since they do not move around in order to stay alive. Through their roots they draw out of the earth what they need in order to live. However, as soon as there is mobility, some warning system is necessary to prevent the danger of collision. The warning could be visual or auditory. If the sense organ is visual, the image may lack all form and merely distinguish between dark and light. The image may also be colorless, as it is with most animals on earth. Perhaps the external sense of Planetarian X, like that of bees, perceives infrared, becoming aware of an encounter through the heat radiated by the encountered object. Or it could be that the inhabitants of Planet X have only *one* sense of a most refined nature susceptible to the slightest stimuli and, like a sensitive radar system, aware of all obstacles within its radius. All of this is, of course, purely speculative, and how the sense apparatus of an unknown higher species has been worked out concretely, we can only guess. Cartoonists have given their imagination free rein on the subject, and it is not our task to compete with their fancy. We would do better to explore that which lies beyond the percept, that which we call intellect.

No one can say with precision what the intellect is, even on earth. It is not in itself a verifiable thing. It is that by which one knows verifiable things or not-so-verifiable things. It defines, yet is itself indefinable. But in claiming that it defines other things, we have already said what it does, and in observing what it *does*, we claim to have some idea of what it *is*. The intellect is that through which we define, in other words, that through which we know. Our hypothesis being what it is, we must attribute that privilege to Planetarians X as well. They *know*—that is, they have an *idea* of something. And being practitioners of Augustine's *cognitio vespertina*, their ideas are obtained through the senses and are not infused.

With the material at hand the Planetarians X make propositions and evaluations. They cannot—nor can we—claim a direct and immediate intuition of all the inferences that lie dormant in a proposition. Any creature in the conquest of truth proceeds like the traveller in the wild who, in crossing a river, jumps from stone to stone. Using "rational discourse," Nhoj, no less than his terrestrial counterpart, proceeds from proposition to proposition, and by the same token is capable of both negative and affirmative propositions.

Living in a world of matter and practicing rational discourse—the two seem to be inextricably interwoven, for if you have the one, the other follows—the inhabitants of Planet X could conceivably use mathematics. Mathematics, proceeding along rigorous lines, is a rational discourse that carries compelling evidence. Although mathematics seems to exist all by itself on a strict theoretical level, free from matter, still it can only be born in a world of quantifiable matter, since it presumes the possibility of "dividing" up the quantifiable and of constructing endless combinations with the resulting units. It seems fair to say that for the same reasons angels or spirits do not practice mathematics. Aquinas would make the point, and rightly so, I believe, that they do not practice rational discourse, since they see all the implications at once. We could say that in some way they are computers of the instant. Angels or spirits, furthermore, exist in a *phusis* beyond matter. That nonquantifiable *phusis* is nonfragmentable; hence it cannot be enumerated or reconstructed in a variety of combinations.

All this does not make it compelling that the Planetarians X are indeed mathematicians. We cannot radically exclude the possibility that their grasp of the external material reality is thorough yet nonmathematical. Even on earth there are a number of people who have striking insights into the nature of things and into certain events, yet whose brilliant intuition rebels against the numerical as the tool for ultimate comprehension. But this fact, which I shall not refute, does not make mathematics invalid as an instrument of universal comprehension and communicability. If the Planetarians *are* mathematicians—and I am inclined to think that they are—we are entitled to say that mathematics is the same on Planet X as on earth, just as it is the same in Washington as in Peking. Obviously we are not discussing here the philosophy of mathematics but merely its content. Consequently, however personal and irreplaceable an angular vision may be, its mathematical content belongs to a realm of universals. As such, angular visions are communicable in toto, so that we might conclude that a full-fledged communication with Planet X could take place along mathematical lines.

One can go a step further and make the point that whoever understands mathematics will uncover the natural laws in physics, and whoever knows natural laws will apply them to the construction of certain devices, as in engineering. There does not seem to be any objection to that sort of deduction, and it seems perfectly logical to let the Planetarians X build the Bridge of San Luis Rey.

We note, however, that as we move away from nonsensory mathematics in a more empirical direction, the incidence of "angularity" increases. A mathematical conclusion is universally valid or invalid. A physical law is universally verifiable. But an engineering accomplishment is already dependent on the skill of the engineer, with the tragic result that the universality of mathematics per se does not prevent bridges from collapsing.

BEYOND THE MATHEMATICAL

In moving beyond the mathematical, we face, instead of universality, the dominant characteristic of mathematics in its general structure, the uncommon or nonuniversal. The uncommon is that which differs from individual to individual and from group to group, both ontologically (in their very being) or noetically (in the domain of knowledge). Hence when we make the point that the knowledge of a certain individual or of a certain group has something *uncommon*, we want to stress the fact that their knowledge has something unique. In addition to the understanding of mathematics that we have presumed to be fully universal, there are forms of knowledge that are what we shall call *uncommon*. How might this mode be fulfilled on Planet X?

We have already set the stage for a noetic differentiation between individuals on Planet X, in considering the existence of a fragmented totality (in space and time) to be more probable than the existence of a massive, symbiotic-type monad. Nhoj, therefore, as part and fragment of the Planetarian X, totality, is in his very being limited and circumscribed and by the same token individualized. As a result of this ontological confinement he is also condemned to his solitary vision: if it is a privilege to be himself, it is also a shame to be only himself. He is caught within the englobing and overwhelming structure *and* vision of the Planetarian X totality and no doubt shaped and molded by it; yet he has his vision, which in the line of our theory will be an *angular* one.

Of major importance for our purpose is the understanding of that which characterizes the knowledge of Planet X as a whole. Although we assert and maintain the uniqueness of Nhoj's vision, it remains a fact that this vision is essentially shaped by the totum to which he belongs. How do Planetarians X approach object A? They do so in line with a mode of approach that is theirs. And that mode is nothing but the result of millions of years of evolution in a different niche (from that of earth) and a different ambience. The inhabitants of Planet X have from the start adapted to a different world, and the millennia of growth and evolution have merely increased the originality of this point of departure.

It is obvious that we cannot be absolutely certain about the sort of idea Planetarians X form of object A. Is it the result of an Aristotelian abstraction by which the common is drawn from multiple concrete, particular objects? Is their "idea" nothing but the diluted image or sense percept, as Hume proposes, for our earthly understanding of it? Is it perhaps a form of intuition not conceptualized the way we know, yet much more penetrating? It would be presumptuous to exclude any of these possibilities and many others. One thing is clear: knowledge on Planet X is *sui generis*. But that sort of observation, which we are compelled to make, undercuts the absolute character of the Greek universal and reveals to us that the primordial feature of all knowledge is its particularity. By this

I mean that it will always be such or so, not just plain knowledge that is the same everywhere. Object A observed on Planet X will become object AX, while on earth it will be known as object AE.

I cannot say with certainty, of course, how object A would look to a terrestrian visitor landing on Planet X, but I am inclined to think that the old adage would still hold: whatever is received is received in the mode of the recipient. Object A, therefore, in the eyes of John, landing on Planet X, would appear as object AE, with the result that John and Nhoj would not know object A in the same way. Nhoj may see a lot more than John; he may also see a lot less. Both may claim that they know what object A is. What they see and know may very well be a definition of object A: They both know *what* it is—and they will both be right—yet their knowledge of object A will not be the same. The definition of object A obtained by Nhoj will be unknown to us terrestrians, just as our grasp of the same object will be hazy to him.

Of course there is no reason why object A could not be known still differently by Totum Z. In fact there is no reason to deny the knowability of object A as being infinite. There is in object A a potential and an overt-ness-to-being-known by an endless variety of collectives in an endless variety of ways. Object A, then, would be viewed and viewed and viewed, again and again and again, differently; yet each one of these viewings would be true. This is where the power of the *angular* truth comes to the fore, since each view of object A is a new discovery and, as such, is unique but true.

If object A is infinitely definable, it does not follow that it *is* everything. Object A can be known as object AX, object AE, object AZ, etc.; yet any one of these "definitions" fits object A and object A only. No mode of knowing object A predicates any attribute of the object that its approach does not allow. If it does, it commits an error. Obviously error is possible in any one of these modes of knowing. Error does take place on Planet X no less than on earth, but the sort of error that takes place on Planet X can only take place on Planet X, for it consists in making an assertion for which this particular mode of approach provides no ground. Let us presume that Nhoj has only two senses; he will come to a definition of object A that is quite different from that formed by John. If he makes a judgment or proposition concerning object A that is not the result of his cognitive approach, yet happens to fit within the angular vision of John, he is in error, although accidentally right in the world of John. However, this accidental truth is catastrophic for Nhoj. His whole approach must fit into a world-picture where object A can only be seen and understood as object AX and not as object AE. To see it as object AE, as John does, would break down the cognitive and pragmatic composition that the individual or the group consciously or unconsciously constructs to make the world in which it lives intelligible and livable.*

At this point one may be seriously concerned with the problem of whether or not John from earth and Nhoj from Planet X can communicate. From the data at hand the answer would seem to be that communication will not be an easy task. So far, it looks as if mathematics is the only common ground. Beyond the mathematical the situation becomes problematical.

However, if there is a consciousness on both sides—as has been our presumption—then it should be possible for John and Nhoj to agree on what to disagree on and to build from there, slowly but persistently, a mode of conversing. So, for example, if both John and Nhoj agree that object A can *truthfully* be understood as object AE *and* as object AX and that on both sides their comprehension of the object is right and valid, we shall rejoice, for great pedagogical progress has been made. Nay, more! A moral victory has been achieved, one that we terrestrians have failed so often to accomplish among ourselves.

One interesting feature should be pointed out: in this complex epistemological problem, faith will play a role. For in seeing that which

---

*Likewise it may not be advisable to destroy the angular vision of a person just because it appears false to one. Not only may that vision make a world intelligible to that particular individual or to the group to which he belongs, but it may also make it livable. I am reminded of the story Mircea Eliade tells in *The Sacred and the Profane* of the group of Australian aborigine nomads who on their wanderings carried a pole with them that they planted in the middle of each new settlement to constitute "the center of the world." When missionaries invaded their territory and took away the pole, as mere superstition, the Australian tribe withered away and died; its members had lost the center of their world. See also my book *The Planetary Man* on "The Sin of the Proselyte" (pp. 313-17), especially: "Such is the defect (and the power) of all proselytism; it is unwilling to see the different and to respect the diverse. Although the intention of the proselytizer is a noble one in many ways, the execution is in most cases a failure because it is the faithful replica of *one* program, *one* idea and *one* mode of action. There is no adaptation, no flexibility but only a fierce defense of the same concept everywhere and always."

I should stress the fact, however, that in the problem here under consideration— the quota of truth and its difference between the two planets—there is more than a mythical and perhaps debatable assertion at stake. It is important to remember that both planets have an angular vision, that each one is right yet not identical.

appears. John can only conclude that object A is object AE. To go beyond this in an attempt to understand what his senses in no way teach him, he will need an assent of the mind not based upon evidence. This is commonly called *faith*. I shall call this the breakdown of evidence as the immediate criterion of truth. However, evidence may still have value as a more remote and mediate criterion. For if I were John, and hence not seeing what Nhoj sees in defining what object A is, I still might lean upon external evidence, the evidence, namely, that Nhoj is a trustworthy witness. But this of course is not direct evidence of what object A really is. The perplexity of John should, it would seem, make one thing clear—the utter naivete of Western man when he thinks that once he has a definition of object A, that is *it*, once and forever.

Indeed, this is the conclusion of this study: We do not know if Planet X exists, we cannot with compelling certainty tell how and what its inhabitants know; but we can conclude, I believe, that the Greek universal, as we have called it, is too shallow an expression of truth to be the only one. Truth is multiple, actually infinite. The collective, whether on Planet X or on Planet Z, carries the burden of knowing the truth. However, it is clear by now that knowledge on Planet X as a whole or on earth as a whole is only *angular* when considered in the total view of knowing in the universe. The notion of a single truth has been useful to science, but when fully observed—that is, including an extraterrestrial range—the same notion loses its firm univocal applicability. The sole universal that could perhaps appear satisfactory and eventually be theoretically aimed at—although in practice never reached—would be the additive universal. Truth, then, in a perfect and ideal way, would be nothing but the sum of the diverse angular truths. No living totum, be it Planet X or earth or any other planet, is sufficiently equipped, by itself, to exhaust the knowable potential hidden in the depth of Reality.

NOTES

1. S. Th. p. L, q50, art. 1 and foll.
2. I examine this elsewhere in a more detailed study which is meant for terrestrial consumption only. See Wilfrid Desan, *The Planetary Man,* Parts I and II (New York: Macmillan, 1972), p. 82 ff.
3. S. Th. p. I, q58, art. 6.

"There are many forms of life in space. Many forms of death, too." So runs a line from a segment of the television series "Space: 1999."

The problems foisted upon us by the fact of death—and all that it means for life—is here examined by Richard Doss, a philosopher/theologian and thanatologist. While having dealt with death as a pastor and friend-in-dying—when he himself listened, often, to the shriek of Ivan Ilyich—Doss's primary feelings about death, as with us all, are personal and existential. "My concern with death is an attempt to come to grips with my own fears, anxieties, and hopes—with my own ephemerality. Thinking of death in this biocosmic perspective has been very significant —and painful—for me."

Dr. Doss is the author of a book on the theology of death entitled *The Last Enemy* (1974), a book that he describes as "a confessional statement at a particular point in my own pilgrimage."

Doss teaches philosophy at Orange Coast College in California.

"'If life can be worthwhile at all, then it can be so even though it be short. And if it is not worthwhile at all, then an eternity of it is simply a nightmare.' If one cannot affirm life to be meaningful at forty, how could he do so at four hundred? Death forces us to deal with the question of meaning now."

Richard Doss

Life and Death in the Biocosmos

In the *Critique of Pure Reason* Immanuel Kant proposed that the entire interest of reason, speculative and practical, should focus on three questions: What can I know? What ought I to do? For what may I hope? Twentieth-century philosophy has responded to Kant's programmatic statement by centering its interest on the first question. The problems of knowledge, especially in the English-speaking world, have captured the thoughts and imaginations of working philosophers. More recently we have witnessed an interest in the second question. Technology and secularization have threatened the values of modern society, and the rapidly growing literature in ethics and values indicates a concentrated effort on the part of philosophers to respond to the perplexing questions that shape the lives of each of us. But Kant's third question—the issues regarding the future—still stands as an open arena of investigation and concern. Indeed, perhaps the major challenge to philosophy in the last decades of the twentieth century is whether it can face the future imaginatively and creatively or whether it will simply be content with a status as a second-order discipline, able only to analyze and evaluate the concepts and ideas of other disciplines.

Philosophical speculations about the future inevitably bring about a reevaluation of our established methodologies and assumptions. Yet this is no new venture for philosophy. The utopias of the past—Plato's *Republic*, Thomas More's *Utopia*, Campanella's *The City of the Sun*, or H. G. Wells' *A Modern Utopia*—all confronted societal assumptions and forced people to think of reality with new models, new conceptual frameworks, in short, a new *Weltanschauung*. A biocosmic perspective invites us to conceptualize in categories closer to fantasy and dreams. Yet we remain creatures tied to space and time. The category of finitude is a fundamental assumption with which every philosophic speculation must deal realistically. To be finite is to have a beginning and an end, and every human being looks at life with the consciousness that it will come to an end. Thus our thoughts about the meaning of human existence are inextricably tied to the reality of death. Mortality and finitude are accepted as basic realities by which our understanding of life is defined.

Our philosophic ancestors did not turn from the motivating power of death. For Plato death was a play for which the life of the philosopher was a daily rehearsal. Plato saw the death of Socrates as the model of the philosophic life, in which death is the fulfillment of the philosopher's true quest. Schopenhauer spoke of the motivating power of death when he said: "Death is the true inspiring genius, or the muse of philosophy. . . . Indeed, without death men could scarcely philosophize."[1] Philosophy for Schopenhauer was a means of overcoming, or at least coping with, the terror of death. In this century Martin Heidegger presents the most penetrating attempt to give a philosophic interpretation of the significance of death. For Heidegger death is an instrument of philosophy, able to lead us into an understanding of Being. Death is portrayed as that which defines the very nature of human existence.

A new era was opened with the advent of nuclear death at Hiroshima and Nagasaki. Death had been singular, punctiliar, the final end of the life of one human being. But nuclear death is corporate, the possible end of the entire human race. Arthur Koestler captures the mood of our own generation when he observes: "The crisis of our time can be summed up in a single sentence: from the dawn of consciousness to the middle of our century man had to live with the prospect of his death as an individual; since Hiroshima, mankind as a whole has had to live with the prospect of its extinction as a biological species."[2] This is the "terror of death" of which Ernest Becker so eloquently speaks in his Pulitzer-prize-winning book.[3] We have lived, in this century, with a conspiracy of silence surrounding death. Death was taboo, not to be mentioned in polite company. Only in the last few years has the silence been broken. A society once known for its denial and avoidance of death is reading and talking about this previously unmentionable aspect of life. Perhaps the realization that death can come so quickly to so many people has forced us to drop the shroud of repression and silence.

We are, sociologists inform us, the first death-free generation. There are many in American society who live to middle age without experiencing death in any first-hand way. Whereas fifty years ago nine out of ten deaths occurred under the age of sixteen, today nine out of ten deaths are of persons over sixty-five. Medical institutions no longer deal primarily with the living, but with the dying.

Death is separated from life by means of the systems and institutions that deal with it. We have turned death over to managers and systems analysts—an inevitable result of a technological culture. Our knowledge of death is thus second-hand. We experience death through television, films, and books, but our experience is detached, remote, and distant. It is *your* death or *their* death that I think about and witness, never *mine*.

In contemporary society death has been stripped of a context of meaning in which it can be interpreted. In secure, rural America death was looked upon as an inevitable consequence of human mortality. Although never welcomed or perceived as a friend, it was interpreted within a religious framework that saw life and death as part of the will and purpose of God. The prospect and hope of a future life beyond the grave eased the anxiety with which one contemplated the future. Today, in urban, technological America, this framework no longer exists. Our contemporary culture has been cut loose from its religious moorings. The idea of interpreting death in terms of a suprasensory, otherworldly realm seems quaint and primitive to those influenced by science and technology. The questions we face regarding identity, dignity, and human values can no longer be answered in terms of a divine plan or purpose. While our forefathers were content to find their hope in the promise of a future life, we want our meaning and hope right now.[4]

Our Zeitgeist is secular, and the process we call secularization has stripped death not only of the context with which it was traditionally interpreted but also of its significance. Secularization has left death denuded, isolated, and separated from life. Thus we tend to look upon death as an accident or a communicable disease. We have been conditioned to think of it as unnatural and uncontrollable and thus something to be feared and avoided. Within a religious structure that deemphasized the significance of life in the here and now it was possible to think of death as a kind of doorway to a future existence. But our secular spirit claims autonomy over our present and our future and is therefore willing to be pacified by belief systems that avoid the reality and power of death itself.

To what does the term *death* refer? Throughout this discussion I assume that death is the final end of finite existence as we now know it and experience it. It is possible to distinguish levels or dimensions of death, such as "bodily" death, "organ" death, "cellular" death, and "social" death; but for our purposes death is the termination of life, the extinction of our finite human existence.

Our difficulty in accepting the finality of death can be seen in a simple analysis of our linguistic utterances. Our talk about death generally occurs in two ways: we speak of our own death as "my death" and of the death of others as "your death." As I sit behind my typewriter thinking about death, it is rather easy to contemplate "your death." I read about the death of others in the morning newspaper; I saw it on the late news on television; it was vividly portrayed in a recent film I saw. Occasionally I see the death of others on the highway as I hurriedly drive by on the other side of the road. Perhaps I am one of those rare persons who has been with someone who died. In such an instant I observe "your death" as the cessation of respiration and heartbeat, in short, the termination of bodily processes. But "my death" is a completely different matter. Not only is it impossible for me to observe, I have great difficulty thinking about it and imagining what it is like. How does one contemplate one's own nonbeing? It was this question that led Freud to postulate that every person, in his unconscious mind, is convinced of his own immortality. In observing his patients Freud discovered that denial functions as a coping mechanism enabling patients to handle their anxiety about their own death. Every time we attempt to contemplate our own death, Freud noted, we do so as spectators, that is, we are unable to think about or imagine our own nonbeing. What Freud described clinically was the overwhelming difficulty we have in thinking about death as a personal experience, death in its interiority. There is a sense in which every time I fall asleep I experience a loss of consciousness that I imagine to be a part of "my death"; in fact, as a loss of consciousness I cannot distinguish between death and sleep. The point is that one's own death evades one's full and complete awareness as an event that will affect one's life and experience.

This discussion opens the possibility of a phenomenological approach to the significance of death. The starting point is to investigate the nature of our experience as we think about and attempt to contemplate our own death. What is involved in my own experience (and I assume your own experience as well)? The first thing I am aware of is ambiguity. When I recognize my own mortality and finitude and the fact that my death is a reality, I feel helpless, even frustrated, that I can do nothing about it. I feel trapped by some power, some force I do not fully understand and cannot avoid. In her clinical experience Elisabeth Kubler-Ross has found that many of her terminally ill patients think of death as an uncontrollable, powerful force that comes upon them and about which they can do nothing.[5] One patient said, "Right now I feel strong and well, but I know I have something growing inside of me that is beyond my control." It is this sense of "no exit," no way out of an existence in which death is the end, that gives rise to the basic ambiguity regarding the human condition.

A second element in my consciousness is anxiety and fear. It would be misleading to assume that anxiety and fear are the same. Psychologists make a helpful distinction between these two experiences. Fear is a common human emotion characterized by the fact that an object must be faced and dealt with. Fear is a psychic response to specific conditions such as height, closed places, or loneliness. Gregory Zilboorg argues that the fear of death is present behind all of our normal functioning as human beings to insure self-preservation. We fear annihilation and extinction. Even though the fear of death does not always show itself in specific ways, Zilboorg contends that it is universally present in every human experience.[6] Anxiety is distinguishable from fear in that it has no specific object —it cannot be faced or dealt with in any concrete way. Anxiety is experienced when we have a vague apprehension that something terrible is going to happen, even though there may be no empirical or experiential basis for such a feeling. Anxiety is an existential awareness of what it means to be finite, that is, an awareness of death as a final, personal experience. We are not anxious about some*thing*—that is precisely the point regarding anxiety. We are anxious about nonbeing, and this ontic anxiety is present in the entire process of living. We cannot escape it.

As I explore my own consciousness regarding death, I find, third, a curiosity, an inquisitiveness regarding the significance of my death. What happens when you die? What is it like to be dead? Do we "experience" being dead? From Plato to present-day writers, we find it common to the human experience to wonder about death. It is in this sense that death is a mystery, a *mysterium* that waits to be unveiled. Some thinkers argue that this kind of wondering about death makes, quite literally, no sense. Wittgenstein once observed, "Death is not an event in life. Death is not lived through."[7] Whether or not we "experience" death or the consideration of death as an empirical reality is precisely my question. The ontic status of death is unknown to us and therefore piques our curiosity. We have a certain awe regarding those persons who face death or have been close to death. Hemingway felt this about the bullfight; our generation of youth turn to an Evel Knievel. We used the term *heroic* to classify those who have faced death with confidence and courage. We find death-seeking behavior dramatized, in auto racing, for example, as a model of heroic behavior. The military model is symbolized by the medals given to those who have acted "above and beyond the call of duty." Ernest Becker notes that we have elevated animal courage into a cult.[8] In ancient times the hero was the individual who went into the spirit world of the dead and emerged reborn or resurrected, having conquered death. Today the hero is one who faces death courageously and does not turn from it.

Another aspect of my awareness is that I look upon death as an intruder, an enemy. We personify death as one lurking in the shadows, moving on the outskirts of life, waiting to come in quickly and without invitation. Most of us have heard the story of the Baghdad merchant who sent his servant out to buy provisions. The servant returned, pale and shaking with fear, and said, "Master, just now in the marketplace I was jostled by a man in the crowd. When I turned about, I saw Death. He stared at me and made a threatening gesture. Lend me your horse and I will ride away to Samarra where Death cannot find me." The merchant lent him his horse, and the servant mounted it and rode off as fast as he could. Troubled by the incident, the merchant went to the marketplace, where he found Death standing in the throng. He approached him and said, "Why did you frighten my servant so when you saw him earlier today?" But Death replied, "I did not intend to frighten him. I was greatly surprised to see him, for you see, tonight I have an appointment with him in Samarra."

Our personification of death is borne out by several recent research studies.[9] A group of college students ranked the following metaphors as appropriate terms for describing death: an understanding doctor, a gay seducer, a grinning butcher, a last adventure, a threatening father, a misty abyss, the end of a song. The women tended to select "a gay seducer"

while the men often picked "a grinning butcher." Not only is the personification of death significant, but it is also noteworthy how the themes of
sex and aggression are linked with death. In a second study, reactions of
women dying of terminal diseases were contrasted with those of a
group of women hospitalized with minor illnesses. The results indicated
that the women with fatal illnesses pictured death as a mysterious stranger
who traps them and forces them into sexual encounters against their will.
A large body of data is now available indicating that death is personified
as an evil intruder who comes to take life from us.

This last aspect of experiencing my death leads to another important
area. Does my experience or my consciousness of death imply that I look
upon it as an evil? In the language of traditional metaphysics does death
have an ontological status? Is there value or loss of value in death?
Throughout his writings Albert Camus wrestles with these questions,
implicitly in the characters of his novels and explicitly in his philosophical
writings. When he writes, "Because of death, human existence has no
meaning. All the crimes that men could commit are nothing in comparison with that fundamental crime which is death," is he not saying that
death is such a noxious evil that it destroys all possibility of seeing human
existence as good or meaningful? I would prefer to say that death itself is
neutral—that is, it has no ontic status—but existentially it is experienced
and dealt with as evil. Death is a phenomenological reality we all must
face. In the evolutionary process on our planet, living organisms have been
developing and mutating for three billion years, yet estimates are that
only 1 percent of the species that have evolved still exist. Life is permeated
with death.

In the history of Western thought it is possible to distinguish two
opposing interpretations of the relationship between life and death: death
as the natural and inevitable termination of finite existence, and death as
that which gives authentic and ultimate meaning to life. Herbert Marcuse
notes that these two poles lead to two contrasting views of life. "On the one
hand the attitude toward death is the stoic or skeptic acceptance of the
inevitable, or even the repression of the thought of death by life; on the
other hand the idealistic glorification of death as that which gives meaning
to life, or as the precondition for the true life of man."[10] Those who look
upon the relationship between life and death in naturalistic or materialistic terms see life as a physiological process and death as the cessation of
biological function. Life and death are nothing more than natural processes. Every person is finite and therefore subject to death as a universal

law of nature. Death is accepted as the termination of human existence.

The opposite pole is expressed in the idealistic interpretation of life as a pilgrimage, a temporary sojourn in which we pass through death to real life. This world and our life in this world is transient and ephemeral, and only death can deliver us from it. This mood is given its first clear expression in Plato and manifests itself in various expressions of philosophical idealism. It is interesting to note the attitude toward life that developed when Platonic idealism was taken into Christian theology. John Calvin, speaking for this life-negating view says, "In comparison with the immortality to come, let us despise this life and long to renounce it."[11] He further claimed that the Christian should not fear death because, after all, the body is an "unstable, depraved, corruptible, frail, withering, and rotten tabernacle."[12] This life-denying posture of Calvin was woven into the fabric of American religious thought, as evidenced in this nineteenth-century hymn: "This world is not my home, I'm just a'passin' through. My treasures are laid up somewhere beyond the blue." This mood prevails in much of popular religion today. It implies a metaphysical framework, utilizing Plato's conception of immortality and a body-soul dualism. Within this framework death is trivialized and denied a place as a part of real life.

Thus the task at hand is to move beyond the materialistic and idealistic interpretations of death that present a death-accepting style. Our attitude must provide a balance in which both life and death can be confronted openly and honestly.

Many cries are heard urging that we deal with death from a new perspective. Alan Harrington passionately argues: "Death is an imposition on the human race, and no longer acceptable. Man has all but lost his ability to accommodate himself to personal extinction; he must now proceed physically to overcome it. In short, to kill death: to put an end to his own mortality as a certain consequence of being born."[13] Harrington's position is quite simple and direct: we must abolish death. Utilizing the tools of medical science and technology, we now see death as a problem for medical engineering, the engineering of our own divinity. Advocating an "immortalist position," he contends that the usefulness of philosophy has come to an end, "because all philosophy teaches accommodation to death and grants it static finality as 'the human condition.'"[14] We must defy death and eliminate from the human situation the inevitability of the aging process. His own interim proposal is cryonics, to "freeze-wait-reanimate" while science looks for and perfects an answer to the problem of death.

Science-fiction writers have presented a vision of endless life. Robert Heinlein's *Time Enough for Love* presents a future in which Lazarus Long not only overcomes the aging process but also can consider the significance of human relationships in terms of time.[15] Lazarus is the perfect specimen of eternal youth, with intelligence, memory, physical and sexual capacity sufficient to fulfill anyone's dreams of a future in which death has been conquered. Heinlein's genius is in his ability to force the reader to feel the frustration of being ephemeral. His insights into the close inter-workings of sex and death are also significant. Death is a final impotence, to be avoided at all costs. The insight of Heinlein is supported by the clinical evaluation of Rollo May: "My viewpoint is that, in human beings, it is not merely the depletion of eros which causes the fear of death . . . but that in all stages of human development the experience of love and death are interwoven. The relationship between death and love is surely clear in the sex act. Every kind of mythology relates the sex act itself to dying."[16]

How do we evaluate these fantasies of a future in which death has been eliminated? What would the quality of life be like for the Lazarus Longs, those who can live forever? Ernest Becker is skeptical regarding what he calls "the one-sided Enlightenment dream." He notes that a postponement of death is not a solution to the problem of death. In response to Harrington's vision, Becker believes that death would be "hyperfetish-ized" as a source of possible danger. If the smallest virus or unexplainable accident could deprive a person not of ninety years but nine hundred, it would be ten times more absurd. Becker likens the utopian views of death offered by Harrington and Heinlein to the beliefs of primitive societies where death was denied as the final end of human existence through belief in life after death. The fallacy of this "sterile utopianism" says Becker is that the fear of death is not the only motive of life. There is also the capacity for heroic transcendence, victory over evil, and the consecration of one's existence to higher meanings—these motives are just as important, Becker claims, as the human nobility gained in facing animal fear.[17] Is Becker simply one more exponent of a philosophy that teaches accommodation, or has he discovered something deep within the psychodynamics of the human species?

Perhaps there is a sense in which a biocosmic perspective can provide a balance between the position of Becker that one "is stuck with his character" and that of Harrington's "new act of faith." Carl Jung said, in

Answer to Job, that a positive response to Job's lamentations lies in recognizing that the human condition will not always remain the same, because a new man will break out of the womb of creation. This "new man" may well be the product of medical engineering, a mutation in the cosmic evolutionary process, or both. Since it seems probable that a life with no terminal point might become tedious and boring, we are aware that the quality of our lives is interwoven with the length of life. Paul Tillich envisioned a "new being." He believed in the emergence of a new kind of human being, one who was more in harmony with nature, less driven, more perceptive, and more in touch with his own creative potential, a being who would move into the future with a vision of communities to replace the collectivities created by our materialistic culture. But Tillich was well aware that the idea of the new being was a myth. He knew it was something to look for and work for but that it could only be affirmed when we have "the courage to be," the ability that enables a person to stand independently and face boldly the contradictions of the real world.[18] To be human is to be open to a future in which new possibilities are close to reality. We must resist the pessimism of a Freud and the explanatory idols of determinism. Rollo May speaks of the capacity to meet anxiety about human finitude in terms similar to Tillich's, as "the courage to be one's self." May views courage from a clinical perspective as "the inner side of growth . . . a constructive way of that becoming one's self which is prior to the power to give one's self."[19] This perspective enables us to look at death not only as an existential evil but as a positive reality that shapes the quality of our lives right now.

The value of life right now is shaped, first of all, by the fact we know it has an end. We may fantasize a *longer* existence but not an *endless* existence. Knowing that life has an end gives a sense of urgency to the priorities and values we set for ourselves. Just as the worth of an object is established by its "preciousness" or novelty, so the value of life is set by our awareness that it is not endless and therefore should not be trivialized or wasted. My death is a constant reminder that life has a terminal point, a boundary forcing the question of life's quality upon me. Heidegger and Jaspers both emphasize this matter of death as a boundary situation: the moment we think that mortality and finitude are outside of human experience as we know it, death confronts us with reality. Our present is interwoven with the fact that life has an end. To Lazarus Long, what would be the significance of "You only go through life once, so get all the gusto you can!"? Death changes our perspective on life.

Another positive value of death is that we live with no timetables, no guarantees as to when the end will come. The simple truth for each of us is that we do not know when we will die. Every day we involve ourselves in situations that could cause our death. Anyone who drives the freeways in and around Los Angeles (or any other city for that matter) is aware of the ever-present possibility of accidental death. Life as we know it has a precarious quality. The husband of a friend of mine stopped off for a drink on his way home from work. While he was there the tavern was robbed, and he was shot and killed. The newspaper and the news broadcast are constant reminders that life is precarious and beyond our control. A dam breaks and several hundred people are killed in the ensuing flood. A plane crashes and a hundred lives are lost. The quality of life is affected by the fact that life is not predestined and is unpredictable. One could only guess at the ways our attitudes might be changed by a greatly lengthened life or even an endless existence. Perhaps we would learn to take our time, to "mellow out" in the knowledge that time is no longer an issue or concern.

We also see that death enables us to evaluate the quality of a person's life as a whole. Death is the final chapter in the book of life, and any critique or appraisal concerning the value of the book cannot be made before it is completed. If you evaluated the life of Adolf Hitler in 1912, when he was an unknown painter of shabby postcards, your appraisal then would be radically different from one made in 1945, after he had masterminded the extermination of six million people. One can only wonder about the appraisal of the life of a Lazarus Long or of a person three hundred years of age with an estimated lifespan of six hundred years. I keep reminding myself that Kant would have been virtually unknown if he had died before the age of fifty-seven, the age at which he published the *Critique of Pure Reason*. On the other hand, it is always humbling to remember that when Mozart was my age he had completed his works and had been dead for six years. The matter of evaluation and appraisal again reminds us that the issues of quantity and quality are interwoven.

All of the significant issues regarding life and death are brought into focus by the question of life's meaning. How we look at life is in large measure shaped by our perspective on death. Camus can say that life is meaningless because of death, but in so doing he makes death the final index of life's meaning. He thereby separates life from death. If we are to affirm that life and death are interdependent realities, we must attempt to deal with the question of meaning without separating one from the other. Kurt Baier takes the position that "death is simply irrelevant. If life can be worthwhile at all, then it can be so even though it be short. And if it is not worthwhile at all, then an eternity of it is simply a nightmare."[20] I suspect that Baier is right. If one cannot affirm life to be meaningful at forty, how could he do so at four hundred? Death forces us to deal with the question of meaning now. It does not permit postponement. The question of life's meaning from a biocosmic perspective may indeed open the larger metaphysical question as to a meaning of the whole of life, but it does not prevent facing and dealing with our own individual meanings in the here and now. We dare not be content with the skepticism that rightly emerges from the epistemic predicament on this earth. We must always be open to the emergence of new possibilities and new meanings. A biocosmic perspective is one starting point in the search for models to provide a philosophy of the future.

NOTES

1. Arthur Schopenhauer, *The World as Will and Idea.* Quoted in Jacques Choron, *Death and Western Thought* (New York: Collier Books, 1963), p. 162.
2. Arthur Koestler, *The Heel of Achilles: Essays, 1968-1973.* Quoted in the *Los Angeles Times*, March 11, 1975.
3. Ernest Becker, *The Denial of Death* (New York: The Free Press, 1973).
4. For a fuller discussion see Richard W. Doss, *The Last Enemy* (New York: Harper & Row, 1974), pp. 1-17.
5. Elisabeth Kubler-Ross, *On Death and Dying* (New York: Macmillan, 1969), p. 2.
6. G. Zilboorg, "Fear of Death," *Psychoanalytic Quarterly* 12 (1943): 465-75.
7. Ludwig Wittgenstein, *Tractatus Logico-Philosophicus* (London: Kegan-Paul, 1922), p. 185.
8. Becker, p. 14.
9. Robert Neale, *The Art of Dying* (New York: Harper & Row, 1973), pp. 109 f.
10. Herbert Marcuse, "The Ideology of Death," in *The Meaning of Death,* ed. Herman Feifel (New York: McGraw-Hill, 1959), p. 64.
11. John Calvin, *Institutes of the Christian Religion,* trans. John Allen (Philadelphia: Westminster Press, 1962), III, ix, 4.
12. Ibid.
13. Alan Harrington, *The Immortalist* (New York: Avon Books, 1969), p. 11.
14. Ibid., p. 25.
15. Robert Heinlein, *Time Enough For Love* (New York: G. P. Putnam's Sons, 1973).
16. Rollo May, *Love and Will* (New York: Norton, 1969), p. 103.

17. Becker, pp. 267-68.

18. Paul Tillich, *The Courage To Be* (New Haven, Conn.: Yale University Press, 1952), pp. 171 ff.

19. Rollo May, *Man's Search For Himself* (New York: Norton, 1953), p. 226.

20. Kurt Baier, "The Meaning of Life," *Twentieth Century Philosophy: The Analytic Tradition,* ed. Morris Weitz (New York: The Free Press, 1966), p. 377.

Death, with all its pain and paradox, is the subject of this science-fiction "memoir" by Herman Tennessen. On a quick run-through, this recounting might appear to be an innocuous tale of adventure with ETIs. But be on guard. The pieces of the story have been deftly fitted together, and Tennessen doesn't spare us: the shriek of Ivan Ilyich, soft at first, builds to a crescendo; and we are brought face to face with the absurd horror of what it means to exist as a death-conscious, Tellurian caddis fly.

Herman Tennessen is Professor of Philosophy and Senior Research Professor at the University of Alberta. He has been a research scholar for UNESCO and other international agencies. He has written numerous books and articles in Norwegian and English, including *Logic, Reality, and History*; *Contemporary Philosophy in Scandinavia*; *The Psychology of Knowing*; and *Inquiries into a Psychological Theory of Knowledge* (with Joseph Royce).

Professor Tennessen is a new philosopher of the old school who delights in satirizing the chic and the fashionable in philosophy, and he does it with the subtlety of a master craftsman.

Are you going to be so serious
About such a mean allowance of breath as life is?
We'll suppose ourselves to be caddis-flies
Who live one day. Do we waste the evening
Commiserating with each other about
The unhygienic conditions of our worm-cases?

Thomas Mendip
in Christopher Fry's The Lady's Not for Burning

Herman Tennessen

Homo Telluris: The Conscious Cosmic Caddis Fly

MY PILGRIMAGE

We had always called it a cabin. But that was really a misnomer. It was much too large. More like an unassuming "villa," I suppose. Built toward the end of World War I for some sixteen engineers who were to have made this their home while planning the construction of a gigantic hydroelectric plant—which, incidentally, never materialized—its two and a half floors had sixteen bedrooms and, in fact, everything necessary for sixteen occupants. Even more remarkable, considering its remote location, which required two days' packhorse travel for a shopping trip to the nearest hamlet, it had electricity and a telephone.

I recalled that as a child it had seemed like venturing upon a veritable expedition when my family decided to spend a couple of weeks toward the end of the summer holidays in this desolate "villa" in the midst of the West-Norwegian mountain range. With my parents, an old aunt, a sample of siblings from both of my parents' earlier marriages, my superstitious "nanny," the obstreperous cook, and a couple of blithesome chambermaids, we took off from Bergen for a long day's journey—a five-hour train ride, more than forty kilometers in three motorcars, ten additional kilometers in a *cortège* of horses, and buggies up to a large lake. There we acquired the assistance of local farmers to row us, with all our goods, to the other side of the lake and to help us with the short climb up to a second, smaller lake, where someone who didn't look too beat would be assigned to run over knaps and moors to fetch the boat from the boathouse beneath the "villa."

Whereupon commenced the most tiringly toilsome, time-consuming task of the whole campaign: to transport the total expeditionary force, with its complete outfit, from one end of the lake to the other by means of a single rowboat. I would be sound asleep in my bed long before the late dinner was announced.

My pilgrimage to this place last year, almost fifty years later, was greatly facilitated by the fact that new roads permitted me to drive my little Passat directly from the Mayor's office in Bergen, where I picked up the cabin keys, to the lower and larger lake. Also, my gear was not so heavy that I could not easily carry it on my back around the first lake or climb with it up to the other. Actually, I did not feel even slightly exhausted when I got there. So, in keeping with the old tradition, I literally ran over knaps and moors. It was for that reason that I found myself quite suddenly confronted with the vista of my sentimental journey's end.

It was early spring. The plants were barely sprouting, the earth was fragrant, and the air was saturated with soundlessness. The lake lay there, all cold, discouraging steel, draped in light-green veils, pallidly reflecting the bleak mountains engirding it. It was as though I immediately sensed something noisome or sinister—I don't rightly know. Perhaps, I reassured myself, it was only the unaccustomed season. In my memory I had treasured an image of this scenery in its slightly melancholic but warm and gaudily colored autumnal splendor: *le printemps d'hiver*. This was still the raw and ruthless, merciless and marblehearted month of March. And somehow the view struck me as less wonderful than weird—eerie, uncanny, incantational—even ominous, I later thought. But perhaps it only seemed so in retrospect, against the backdrop of the ensuing, harrowing experience.

For it was that evening that I found Hilary Finley von Habermann. Or rather: his mortal and literary remains.

The widow may safely discontinue her monthly advertisement offering rewards for information as to her husband's whereabouts. Her husband is dead. I buried him myself. He lies in a shallow, unmarked grave in the midst of a mountainous wilderness. I know that he would have wanted it so, and I sincerely hope that this disclosure is not going to incite some relative or other to attempt to locate his body and bring it back to Hilary Hall, near Holyhead in Anglesey, from where he, according to the newspaper reports, vanished the previous year. He was last seen, it was said, at about four o'clock on the morning of June 24, clad in some grey sportswear, bareheaded, with a cane, and with a small knapsack on his back. He was, apparently, heading northeast, into the mountains.

He had been dead for almost half a year when I discovered him. When I turned on the lights in an upstairs bedroom I saw him: sitting upright on a simple wooden chair, his neck resting on the top of its back; his head was awkwardly tilted backward, and both forearms rested on a primitive desk, made of a door placed over two chests of drawers.

The climate is damp in western Norway. The stench in the room was unbearable. I scrambled down the stairs, my stomach filling my mouth to bursting. I threw up into the bushes.

It was dark now. And cold. No moon. Only the pale light from that upstairs window.

My first thought was: Thank God there's a telephone in the house! I can call the authorities and leave the matter to someone less sensitive, more thick-skinned and accustomed to handling such situations. With the new roads and all, they should be here tomorrow.

Tomorrow! What about tonight? I sat there for a while, shivering. Then—I shall never know how, or even why, I mustered up the courage—in a peculiar mood of distraction I returned to the house. Slowly I went from room to room, turning on lights and the heat and opening all the windows. The last ones, those in von Habermann's study, were already open. I must have opened them myself, I suppose, when I first entered. From the cellar I had brought with me all I could carry of old newspapers. I covered the floor to the left of the chair in which the corpse still sat. Gradually, I tilted the chair to the left, permitting its contents to slowly slide onto the papers. I rolled the papers around it, roped the whole thing into an anonymous package, and then pushed it into a plastic garbage bag. The frost had not entirely released the soil, but the digging of the grave was not difficult at all. The worst was already over.

Strangely enough, it never for a moment occurred to me that what I had just done might not have been proper. I was simply relieved, nearly euphoric, in fact. I am not sure, but I almost believe I was singing on my way back.

The house was cold. I closed the windows, checked the electric heaters, turned off the unnecessary lights. Again the last room on my route was that "study." There was scarcely a whiff left of unpleasant odor. I closed the windows, and already had my hand on the light switch when I discovered a manuscript on the improvised desk. I had, of course, noticed that small pile of papers there before: long, yellow folio sheets, neatly stacked. But now what caught and retained my attention was the title page. In neat, finickingly embroidered Gothic calligraphy it read:

Ueber den weitreichenden und tiefgehenden Einfluss der Sterblichkeit auf das Leben und das philosophisch—wissenschaftliche Wirken irdischer Menschen, und wie er durch meinen Verkehr mit den ueberirdischen Wesen des Planeten Zeteticas, enthuellet geworden wurde. Bekenntnisse eines noch-nicht-gestorbene Sterbliches, Hilary Finley von Habermann.

Fascinating! Was it merely another science-fiction story? Or, as it were, the deathbed confessions of a space traveler who, as the title said, through intellectual intercourse with inhabitants of a distant planet had come to see terrestrial existence in a different light? I had to find out. I was, after all, a philosopher. In fact, I almost slid into the chair. But then, with a shiver, I controlled myself, collected carefully the whole stack of folio sheets and brought it down to the dining table on the long glassed-in veranda that ran along the south-east side of the house. I brought in an extra lamp and commenced reading. Von Habermann's German was old-fashioned. Sentences sometimes ran on for over several pages, with all the verbs at the end. But having been forced to read Wundt and Ebbinghaus during my period of study in Marburg, the complex and cumbersome style presented no forbidding obstruction, and the content was indeed entrancing. Dead tired as I was, it kept me so spellbound that as I turned the last sheet of paper, the sun exploded over the mountains with a blindingly bright morning. It was April first: fool's day! But I had not been fooled. Whether von Habermann's fantastic tale was fact or fiction seemed almost irrelevant. One thing was quite unquestionable: It had in some insidious way cryptically twisted my most fundamental presuppositions and smashed the whole framework within which I had hitherto so comfortably appraised all aspects of human life and endeavor—including my own.

What had happened? Frankly, I am not sure I can answer that

question. Let me rather make an attempt at retelling the major points of von Habermann's sometimes almost indecipherable but stupendous story.

FROM THE MEMORABILIA
OF HILARY FINLEY VON HABERMANN

ENCOUNTER WITH THE ZETETICANS*

I am German [wrote von Habermann]. My family resided in the little town of Krümmel, not far from Hamburg, where my father had semiretired after a long and extremely profitable business career. He was already sixty-six when he married my English mother, née Finley. She was visiting with her sister—whose husband was David Hartley, Jr.—at Hilary Hall in Anglesey when I was born on August 3, 1797—more than a month too early, but still weighing twelve pounds. It killed my mother. My father left me for a while in the care of Uncle David and Aunt Henny. They were childless and spoiled me quite awfully. So, incidentally, did my father, who took me traveling all over the world. But Hilary Hall remained my second home. And that is where I first became interested in philosophy, through the writings of David Hartley, Sr., a peculiarly pessimistic "materialist." My father was more than delighted to hear that I had decided to devote my life to philosophy. But philosophy, needless to say, could only be properly studied at a German university. "English," he used to say, "is a language for businessmen and drunken sailors."

Intrigued by Fichte's *Wissenschaftslehre* ("theory of science"), I went to the recently established university in Berlin to study under him. It turned out to be a most fortunate choice. *Everybody* seemed to come to Berlin: Schelling, Schleiermacher, and Hegel. I even made the acquaintance of students very much my junior who were later to make a name for themselves, such as Ludwig Feuerbach and Kaspar Schmidt. But most of all I found myself in tune with the philosophy of Arthur Schopenhauer. When he, after an unsuccessful attempt at competing with the great Hegel, left Berlin for Frankfurt am Main, I left with him.

I maintained at the same time my original interest in the relationship between science and philosophy. And when Charles Darwin published his *Origin of Species* in 1859, I decided, at the rather advanced age of

*Von Habermann's introduction to his diary was undated but probably written between July 24 and August 3, 1975.

sixty-two, to throw a bit of color into my unspectacular career by offering the first German translation of this, to my mind, earthshaking work.

My father, as well as Uncle David and Aunt Henny, had long since died, my father leaving me with an income as comfortable as Arthur Schopenhauer's or Sören Kierkegaard's (another acquaintance from Schelling's lectures in Berlin) and my Aunt Henny (who survived her husband by a dozen years) leaving me Hilary Hall and a cabin in the Norwegian mountains. The cabin, however, was strictly for salmon fishing, which had always seemed to me an incredibly silly "sport." On the other hand, I learned to enjoy hiking in the mountains. And one lovely evening in late June the year my father died, 1835, I discovered this delightful, lonesome little mountain dairy farm. I was welcomed with such sincere heartiness— it was as though I were a long-lost relative—that I made a point of coming back every year. They set off one of their tiny huts for my use, and it was there, I decided, that Darwin was to be translated.

There is no night in Norway in June. There is an almost direct transition from dusk to dawn. It was during this brief interlude, when I had my candles burning, that the door quietly opened. I turned, startled, and found that three individuals were standing inside the room by the door. They were about my size or a bit smaller, wearing a most peculiar one-piece habit. Although they had well-proportioned, normal heads, their facial features were strikingly tiny and their ears low seated. Nevertheless they managed to radiate a sort of reassuring, benevolent disinterest, as if they had said to themselves, "To what better use can we put our tiny faces than to let them express a bit of kindness while we are thinking of something else."

They carried small, square boxes on their chests about the size of a modern hearing aid. A beep came from the box of the figure in the middle. And then, in German: "In which language do you prefer to communicate? German, English, or Norwegian?"

"German," I replied automatically.

"Good," said the box in the middle. "Very good." And the mouth, from which no sound had come, opened to a silent smile that revealed a pearly string of tiny teeth. "Then permit us to explain our presence here and our business. We are, you might say, philosophers, like yourself. We come from a satellite, a kind of planet, if you want, presently circling the star your astronomers refer to as Alpha Centauri. We occasionally have visited your planet to observe the evolution of your species. Now you seem to have gained some insight into your own evolution. Some of our researchers in cosmo-ethnology have advanced a series of conflicting guesses as to what you would do when you reached such understanding of your origin and probable destiny. Some have foreseen mass suicide. Others think you might abstain from propagating. Apparently nothing like this seems to have happened."

I had, through all this, remained stunned and speechless. For a while I was convinced I was either dreaming or hallucinating, but then I found myself irresistably drawn into a conversation with that "box."

"Surely," I heard myself say in a tone of dejected irritation, "you must realize it is much too early to expect any consequences whatever from Darwin's work, just a few months after its publication! I am sitting here trying to translate it into German . . . "

"We understand, of course," came a voice from the box of the individual to the left. "We thought, however, that the avant-garde would have known for some time, and that they would have demonstrated *some* of the anticipated symptoms. Look at yourself, for instance. You know as well as you can know anything that in a short while—twenty, thirty, fifty years, for the difference it makes—you will surely be dead. And that's it. What strange psychological mechanisms permit you to live your allotted span of time as though it were some sort of eternity and then, when time is up, to lie down, like your salmon after spawning, and accept death?"

"My favorite teacher, Dr. Schopenhauer, has given much thought to that problem . . . " But there was no need for me to continue. They knew all about Schopenhauer, it seemed.

"He believes," spoke the middle box, "that without death there would be no philosophizing. We are, at least by your standards, immortal; and we have devoted the indefinite tenure of our lives entirely to an activity and a concern that I am quite sure you would be inclined to deem philosophical, were you to understand what it is. Moreover, we think Schopenhauer is too old, too rigid, and too far gone. He is in an almost irreparable state of deterioriation. We give him, say, half a year. "

I had not the faintest notion what they were talking about. Too old? Too far gone? *For what?* I wanted to know.

There was, as I recall, a fairly long silence, finally broken by the figure to my right. His box spoke in the same mild, monotonous manner. But what he said to me was so strikingly extraordinary that it forced me to reconsider my earlier assumption that I was dreaming or hallucinating.

No one tried to stop me as I rushed out the door. I looked around me. It was all there: a bleak moon in the brightening sky; gleams of dawn light washing over the mountaintops; a silvery rippled lake; the pale, peaceful pastureland around the huts, with silently ruminating cows, only adding to the interminable tumbling of the great gray quietude with an occasional long and lazy low.

This was no dream, no hallucination. Suddenly from the chickenyard sounded the pickax voice of a rooster trying to break up a night that never really was. And as if this were a signal, the three strangers emerged from my hut. They said nothing. No sound came from their boxes. But I realized: they were waiting for my decision.

In retrospect, I do not quite understand why I was so stunned . . . or so hesitant. What they offered me was, if not eternal youth, at least some sort of eternal existence. There was nothing sinister or Faustian in the deal. Why should I not accept? I had no family, no close friends. All they asked was that I join them in their journey back to their cosmic residence. As soon as I arrived there the aging process would reverse, or at least halt indefinitely. This would permit them to *study* me, they explained. *How* and *why* they would "try to make intelligible" to me en route to Zetetica. And should I, against all their expectations, at any time wish to *return* to my home planet, they promised to comply, though they considered such a request unlikely.

How could I lose? On the other hand, the whole thing seemed totally unreal. My dear reader—should I ever have one—you must understand that the final proof of the Copernican system had been established only some twenty years earlier—in 1838, in fact—when the parallax of 61 Cygni was measured by my old friend Friedrich Wilhelm Besel, and that of Alpha Centauri the following year by, I believe, Thomas Henderson. And I was to be on my way to a planet of sorts encircling the very same star whose parallactic angle Mr. Henderson had so recently determined! But, I *was* on my way, still in a daze after having accepted the Zeteticans' invitation. I was on my way, literally speaking, out of this world.

August 3, 1975
THE ZETETICAN CONCERN WITH HOMO EPHEMERALIS

Please accept my apologies. I have always had inclinations toward the dramatic. Or maybe I merely wanted to intrigue you, to entice you into delving into my chronicle. But as you now realize, I *did* return from the stars; and I shall try in what follows to guard against my high-flying Thespian tendencies and make my notes more, as it were, down to earth.

Today is my birthday, both according to my *factual* certificate—
which makes me a hundred and seventy-eight—as well as the faked
records, which give 1916 as the year of my birth. And here I sit, over-
looking the same lake from which I departed to Zetetica a little more than
a hundred and fourteen years ago on an early midsummer night's morning
in 1859.

I remember it well.

It is a pity, I suppose, that my interest in science has been purely
theoretical. I have absolutely no sense of, or aptitude for, technological and
mechanical devices. Consequently I am at a loss to explain how the contri-
vance worked that took us from the lake to dock up with a larger vehicle
hovering above the earth's atmosphere.

I was received aboard with the same expression of distrait benevo-
lence. There were only a few with whom I could communicate, and they
were easily recognized by the hearing-and-speaking aids attached to their
chests. Even with them the interchanges were rather limited, apparently
because of the difficulty of "popularizing" answers to my questions to
make them intelligible to me. At times I felt humiliated. When a few years
after my Tellurian return I read about scientists communicating with por-
poises, I felt that I understood the problem the Zeteticans faced with regard to some of *my* questions. What would the scientists have done if they
were confronted with the problem of explaining to a porpoise the more
intricate gadgets inside a submarine! To my query, for example, about
why we did not experience a state of weightlessness on our voyage, they
told me only that they had somehow utilized the acceleration and deceleration to produce a simulated gravitational field. That, I said to myself, I
could almost comprehend. But, I am bound to admit, most of their explanations—when they attempted to give them—were over my head. I am
ashamed to admit how little I learned about their technology from my
eighty-six years on Zetetica.

We landed not on the planet itself but on one of its thirty-six artificial
"moons," from which there was a continual shuttle to the planet's interior.
Yes: *interior.* All the stations were subterranean. Zetetica had been geo-
logically dead a good milliard years before the Zeteticans had noticed the
slow swelling of the aging star to whose system their planet originally
belonged and had been forced to go scouting for a "younger" sun. The
division of work had been trinal. Expeditionary forces were sent out to
investigate solar systems that their astrophysicists believed most likely to
have hospitable planets. Others worked on accelerating the velocity and ex-

panding the orbit of Zetetica and thus preparing it for "takeoff." In the meantime the major work was going on inside Zetetica, trying to make the interior of the planet livable for its inhabitants during the long intergalactic voyage. Our own sun had been high on the astrophysicists' list of candidates for a new solar home, and the scouting parties had, they admitted, been quite intrigued by the external resemblance to Zeteticans shown by the species inhabiting the third planet, *Homo sapiens ephemeralis.* On the other hand, they found our system planetarily overcrowded. So they finally settled for the nearby planet-free star, Alpha Centauri. From there it would also be fairly easy to follow the development of a species that appeared to be at an evolutionary stage not very dissimilar to one the Zeteticans presumably had passed through themselves before their recorded history: *a stage where an individual's tenure in life was fixed within rigid limits, while he, at the same time, had reached a point of some degree of awareness of his grim conditions but not enough total insight to permit him to alter them significantly.*

It was commonly agreed among Zetetican scientists that this stage of ephemerality-cum-awareness must have lasted for several millennia.* After a considerable period of time their researchers became deeply concerned with psychological explanations of how their species had managed to endure this condition, which, in the minds of the present immortal Zeteticans, appeared as an insufferable evolutionary epoch.

This, as they explained to me, was the main reason they had brought me along for more thorough scrutiny. To them I was a live fossil of predominantly paleontological interest. But, as I would later realize, not solely so. . . .

August 6, 1975
WHAT I LEARNED FROM THE ZETETICANS

The mosquitos are no bother to me any longer. I followed the example of the Samic population in northern Norway: I let myself be thoroughly stung the first weeks I was here. Actually it made me quite ill. But now, at last, I am immune. I spent all day yesterday collecting blueberries. Blueberries and trout—that is my entire menu.

I came here to die—if I must. Or, hoping against hope, perhaps to be "saved" by the Zeteticans. After all, it was here that they first found me.

*A Zetetican year is very nearly the same as a Tellurian, as I shall later explain.

And when I, in my ignorance, decided to return to earth, they left me on the other side of this very lake on May 18, 1945—some time, I thought, before the farmers from the valleys were to bring their cattle to their summer dairy farm. Now, thirty years later, I discover to my dismay that the Norwegians had long abandoned their mountain farming. But then I was not aware of that. In a few weeks, I thought, I could start listening for the cowbells and cattle calls.

I found my hut virtually unchanged after eighty-six years. And, most incredible of all, in a rather obvious hiding place behind a loose rock in the foundation wall, I found a considerable sum of now worthless old German marks, plus some thirty-six hundred English pounds in sovereigns and banknotes, which, strangely enough, I had hurriedly stashed away the morning of my departure from earth. A bulky bag with some large silver coins—what the Norwegians called *specie daler*—I had left for my hosts. But, all in all, I must have had a suspicion that I would once again return to this place.

Not that I needed the money. My Zetetican friends had left me with a generous but ludicrously unmanageable farewell gift of 720 gold bullions. Absurd, of course. I laughed in my bed, peeping through the half-open door at the neatly arranged, outrageous pile of shining bars.

It was well into the next day when I first discovered this mysterious monstrosity of a house on the southwest side of the lake a couple of hundred meters up from the shore. I also noticed a boathouse. The house was deserted, and the boat leaked. But the wood swelled to stop the leakage during my fourth or fifth trip to ferry gold and gear across the lake. I carelessly stowed away the gold in the unlocked root cellar and then climbed up along the pillars to the large partly built-in veranda. Then I opened the front door from the inside.

I can still amuse myself today reminiscing about all this—my very first encounter with electricity and telephone. My Zetetican friends would undoubtedly have found both to be interestingly primitive. I marveled at them. In fact I improved my technological competence more, I think, in the few months after my terrestrial return than during all the years on Zetetica. They were too far ahead of me there to be reached by my capacity for comprehension.

Nevertheless, a few things I gathered, of course—some of immeasurable significance, others merely interesting.

First I gradually realized what I suppose I had always suspected

—and this was the most important lesson I learned during my sojourn: I came to see that *a life with a definite, limited tenure was a life neither worth living nor worthy of being lived.* I saw through all the presumptuous pretexts and silly subterfuges of ephemeral man's musing reveries. On the other hand, I was bound to ask: If I understood all this so well, how could I account for the fact that I now found myself back here again, stuck on earth and knowing equally well that there is nothing I could do about it?

There is, I think, an explanation of sorts.

Roughly and generally indicated it is this: Our self-identity is, needless to say, no stable entity. It changes: it develops. But this evolution is not a smooth, continuous unfolding of a 'self.' It proceeds in leaps and bounces from one stage to another. We, as it were, *shed* one 'self' for another. Speaking for my 'self,' I remember when it dawned upon me—I may have been only ten years old—that one day I should no longer be a 'self' that could—or *would*—play Robin Hood and the Sheriff of Nottingham! I shuddered at the thought. And yet the 'self' that was I after less than another decade, would smile embarrassed at any reluctance to abandon the Robin Hood-and-the-Sheriff-of-Nottingham 'self' that I once had referred to as "myself." It was a different thing with the 'self' that was transported to Zetetica. After some fifty years or so, I finally found myself forced to withdraw from the prospect of ever graduating to a 'self' that would be fully capable of playing, as it were, the Zetetican game.

They were all "nice" enough to me. Although I occasionally found their ineradicable expression of distrait kindness somewhat unnerving, they never treated me merely as an interesting exemplar of an extinct species. They did make attempts to educate me—"trying to make matters intelligible to me," as they said. Among the least puzzling pieces of information that I managed to comprehend were the following:

Zetetica is only slightly smaller than earth. It moves around Alpha Centauri in a near-perfect circle (which, incidentally, violates one of Kepler's "laws"). It is 120 million kilometers from its sun, which it takes exactly 360 days to circle. Every day is divided into twelve intervals (rather than twenty-four hours), which are subdivided into 120 intervals, or, if you wish, minutes. All these numbers, however, are "round" numbers in the Zetetican number system, which is, if I have understood it correctly, duodecadic rather than decadic. In other words, it has two digits to which our Tellurian number system has no equivalents. The Zeteticans consider themselves extremely fortunate in this respect since the duodecadic number system is in many respects superior to the decadic. They did not "invent" their duodecadic system any more than we can be said to have invented our decadic one. The difference can be traced back to a sheer coincidence in the great evolutionary tombola: Tellurian vertebrates happened to find themselves with pentadactylic extremities (that is, five fingers or five toes on each), whereas our Zetetican counterparts were lucky enough to have hexadactylic extremities (with six fingers). Hence their *duo*decadic number system.

Another intriguing characteristic of the Zeteticans was their peculiar combination of dispassionate, latitudinarian open-mindedness and their scrupulously heedful and considerate conservatism. Or, perhaps, the now so a la mode "conservationism" would be a more appropriate term.

I shall probably never understand how they sustained and preserved their apparently perpetual survivance and vivification. The Zeteticans did seem to eat and drink, and quite regularly so, but, as they explained, not in order to take in nourishment of any kind but simply to keep the intestines and the whole metabolic system in *preparedness* for an emergency. They even had "exercising machines" to prevent muscular atrophy— "just in case." It was for the same reason that they kept a rich assortment of fauna and flora on the parklike surface of the planet. To maintain preparedness for an emergency was also the main reason they were orbiting a star in spite of the fact that Zetetica could exist as a community quite independent of the solar energy squandered by some sun. The Zeteticans seemed to believe that their relative success as a biological species was primarily due to their reluctance to participate in an evolutionary *gamble*, as other animals had done. They had never staked their entire evolutionary future on developing wings or fins or on any other irrevocable specialization. They had meticulously refrained from evolutionary commitments that could not be reversed, and they were very definitely set on continuing that policy. The only exemption, as far as I could see, was made for the development of the brain (including the eyes), and the cortex in particular. Zeteticans, like us Tellurians, are essentially brain/eye animals. But unlike us, they have in some significant sense freed themselves of their past and made their *ratiocination possibilities* independent of their biological and neurological equipment. For them, literally speaking, *anything is possible*. I do not quite know how to put this, but let me say only that had the Zeteticans, in any Tellurian-like sense of "language," had languages, they would be languages in which anything could be said and understood. Or, expressed in a more chic philosophical fashion: Their "linguistic performance" and "communicative competence," or whatnot, were unconditionally independent of, and totally uninfluenced by, any "deeper" determinants, for example, the neurobiological edifice of either the single individual or the Zeteticans as such. (Needless to say, there would be no concern with what some Tellurian linguists a few decades ago referred to as the "deep structure" of language.)

But then again, in what sense of "language" could there properly be said to be language on Zetetica? I am no linguist, of course, except in the old-fashioned sense of mastering quite a few languages. One thing, however, was to me indisputably clear: Zeteticans never made any noises at one another for *communication purposes*. They did exercise their vocal chords at predetermined intervals—a solitary, conservational-conversational exercise. As for writing or printing, they did nothing of the sort.

Yet, there were, in a sense, languages on Zetetica—many languages.

But at this point I really must apologize. My reader, whoever you may be, please bear with me during the following paragraphs. Not only am I nearly senile, and rapidly deteriorating, I am not at all sure I *ever*, even at my very best, approached the point of grasping the essentials of these most significant aspects of the Zetetican *Lebenswelt*.

You may recall my mentioning that I first went to Berlin to study philosophy of science with Fichte and then proceeded to study with Hegel and Schelling. And, while still remaining a passionate adherent of Schopenhauer, I could not help admiring the Hegelian system and comparing it to those of Aristotle and Spinoza. Sometimes, it seemed to me, Hegel was Spinoza set in motion. But the problem here is obvious: *How can one possibly pretend to compare and evaluate global systems?* Except, it seems, in a mindless mood of shallow, deprecative chatter and small talk. If we were, on the other hand, to be serious and "rational" about comparing and evaluating systems, where would we look for the criteria, the scales, the measuring sticks? Needless to say, we would have to look *nowhere but within those very global systems*, as I shall later discuss in some detail.

The "languages" on Zetetica, it seemed to me, were *like* such global systems. There is nothing comparable to a *translation* from one Zetetican "language" to another, no interlinguistic communicability. Maybe it would be less misleading to refer to the Zetetican "languages" as "meaning-universes," "conceptual frames of reference," or "worlds"—W_1, W_2—W_{n-1}, W_n—where *world* is used as in "the world of theater," "the world of crime," "the world of sport," and so on. I was led to understand that it would be possible for an individual Zetetican to be, as it were, "at home" in an indefinite number of "worlds"—although not simultaneously, of

course; the *number* of worlds with which a Zetetican could be familiar or, if you wish, could master, varied widely from one individual to another. In point of fact, this was the only important respect in which the Zeteticans seemed to divaricate. Such primitive differences regarding color or sex, for example, were, needless to say, unknown. But when it came to numbers of "world-familiarities," there was clearly a form of labor distribution. Some —a minority, I gathered—would specialize in one "world," say, W_i , with which they would make themselves thoroughly familiar. They would generate indeterminate sets of explanatory systems, proliferate and pro-lificate theories, and churn out a lavish array of hypotheses. Then they would amass a myriad of something that, for them, seemed to play a role within W_i not too different from what, within a naive-realistic Tellurian frame of reference, would be ascribed to the sort of "protocol sentences" we traditionally refer to as stating "evidence" ("sense-data," "facts," "observations," etc.). And just as the case would be with our evidence, data, or facts, they were sorted out in types, according to whether they were to be listed as predominantly confirmatory or disconfirmatory instances in relation to a given hypothesis, H_i (within a theoretical frame-work of an explanatory system in a "world" W_i). In addition, confirma-tory and disconfirmatory instances were also arranged along a continuum, indicating the *degree* to which they either strengthened or weakened H_i. However, there was one striking difference: These apparent counterparts to 'facts' or 'evidence' were invented or constructed rather than found, dis-covered, or inadvertently "stumbled over" the way they often seem to be within the thought patterns of contemporary, more or less naive realist-empiricist scientists.

As I have said, this is all extremely difficult. Every now and then I find myself forced out of my chair, going out for a stroll to collect my thoughts and try to think over these problems. That is why it takes me so long to write. My head is peculiarly light. I am weak. When the time comes, there will be little left of me to bury, I am sure. But first I *must* get all this written down. And yet, I am not even sure I understand exactly what it is that I am trying to explain. I was never truly sure when I was at my best, let alone in my present debilitated condition. All I know is that there were on Zetetica an undetermined multitude of *something* along the line of what quite a few post-World War II philosophers and scientists have tended to talk about in terms of "conceptual schemes," "frames of re-ference," "meaning-universes," or the like, except that on Zetetica they were all global and totally and completely consistent. I began by thinking of them as entirely different *languages*. But if that were to be considered the appropriate comparison, we should have to think of languages so entirely different that to change from one to another it would be necessary, in Tellurian terminology, to change *everything* concerning, say, perceiving, feeling, evaluating, conceiving, cognizing—that is, *everything* that could possibly be pertinent to making any kind of judgment within any field of endeavor.

Perhaps I could intimate at least a possible starting point for a direction of thought here if I were to draw attention to the now commonly accepted notion that the cognitive content of facts, data, observation claims, etc., is bound to be a rather immediate function of a conceptual framework tacitly assumed by those very same explanatory, theoretical claims for which the observations were advanced in order to furnish confirmatory instances or strengthening evidence. This is strikingly clear in the example of naively observed light points in the night sky. What is seen or observed will necessarily vary in correlation with the observer's cosmological perspective: a ptolemaically oriented observer's perceptions of the light points is no doubt bound to be entirely different from that of a post-Copernican observer's. *An uneducated child and a trained astronomer, both relying on the naked eye and their twenty-twenty vision, will literally see a different sky.* It has often been argued that if Kepler and Brahe decided to go out one clear and pleasant evening to admire together the same sunset, they would surely not have seen the same thing if each of them had his different cosmology saliently present. Brahe, and he alone, would actually see the sun go down. Kepler, on the other hand, would slowly lose sight of the sun as the earth turned. Were he ever to use the phrase "the sun goes down" it would be in a superficial, semantically unambitious, quotidian vernacular, or the expression would be employed poetically and metaphorically.

On Zetetica, however, where systems are all-embracingly global and entirely clear and consistent, there would be no possibility at all for direct communication between, say, a W_i and a W_j research specialist. Therefore they need W_{ij} researchers; and W_{ijk} researchers, W_{hijkl} researchers, and so on. The higher the number of "worlds" a researcher could master, the more superficial his familiarity with each single "world," but also the broader his "understandings horizon." These, as it were, "multilinguistic," "polyglotic" oriented researchers seemed to serve a function something like that of inspiring popularizers. I therefore asked for permission to communicate with someone who had familiarized himself with *all* of the "worlds." That, however, was out of the question. I learned, first of all, that there cannot really be a definite number of "worlds." Second, no single Zetetican would master more than twelve "worlds"—each's mastery overlapping that of others. Only the "thinking machines," as they were called, could handle more than twelve worlds. And even within their ranks, there was a hierarchy. Only one machine had access to all, at any given time, recorded "worlds" and that was "the great thinking machine" in the Zetetican interior, possibly some sort of computer, I should think. It was divulged to me that it was through this major machine that all my communication with Zeteticans had been mediated.

I am tired. I could fall asleep on this wooden chair with my neck resting on the top of its back. But if I do, will I ever awaken again? I have so much more to say. . . .

There was another remarkable aspect of the Zeteticans' research procedure that I must not fail to discuss.

In my early nineteenth-century naiveté I once asked my hosts how the Zeteticans would verify or refute theories or hypotheses. I had to wait a while for their answer. Then it came: *Such obsessionlike fascinations with definitive refutation and verification of hypotheses and conjectures of any kind were the exclusive and differentiate earmark of ephemeral beings. Because of the alarming limitation of their life-spans, ephemeral beings were panic stricken into a desperate demand for final, conclusive, unmistakable—or at least lasting-looking—"results" preferably to be obtained within the frames of the individual researcher's already narrowly restricted existence.*

"With us"—the voice came from a box that happened to be affixed to one of the first Zeteticans I had encountered—"with us all this is entirely different: Irrespective of the "world" (W_1) or "worlds" (W_1, W_2, W_3, . . . W_{n-1}, W_n) with which we have familiarized ourselves, all our researchers have one thing in common: We are almost literally without limitations of space, time, or resources. The nearest research object of any importance is more than eighteen milliard light-years away. Of the possible cosmic catastrophes worthy of our researchers' concern, none is less than nine billion years in the future. That would give us ample time to prevent, or at least prepare, to survive a catastrophe were one ever to occur. Under such circumstances, you will understand, we cannot afford to completely renounce and abandon any results of meticulous brainwork, whether it is produced by a team of Zetetican researchers or by one or more of our thinking machines. Hypotheses, theories, explanatory systems, etc., that do not at one time, t, look promising within a "world", W_i , may occasionally be *shelved*, but never simply just discarded. Who knows, there may come another time, t', within W_i where they may again prove significantly useful. And there are also other possibilities. But I am afraid we shall have difficulty trying to make them intelligible to you." (The usual refrain, I thought.)

September 1, 1975
ARRIÈRE PENSÉES CONCERNING THE SO-CALLED WORLDS

I have written nothing for several weeks now. My last entry was a major effort for me, and an important one at that, I dare say. The division of the Zetetican *epistémè* into different "*worlds*," or whatever, has been much on my mind in the past weeks. I still cannot pretend that I understand it all. What particularly boggles my mind is the very idea of a multifarious bevy of incommensurate, autarchic, global conceptual systems, meaning-universes, or "worlds." There seems to be an almost paradoxical conundrum here, at least within my "world"—insofar as it even makes any sense within my "world" to talk in such a way as to suggest that one "*has a world.*" And the paradox—if that is what it is—I can only preparatively state as follows:

On one hand, I realize the necessity of a precise and consistent framework in order to make differentiative judgments—either appreciative or disclaiming—with regard to any phenomenon that appears to my consciousness. Let us say, for instance, that one of my learned friends alleges to have made an earthshaking discovery. He has "found" some startling "scientific fact," or whatnot. *I can only engage in what in my days would be called a* rational rencontre *with my friend provided I can meet his claim with a clear and distinct cognitive latticework, with a fabric, a texture, fine enough to fix on his "fact" and trace its possible impact on the whole structurer's putative foundations, as well as the effect it ought to have—according to the system—on my currently most pressing priorities* of any kind: action, intention, volition, cognition, perception, emotion, evaluation, or whatever. Otherwise the effect would be an irrational or "mysterious" one, in the sense that it would merely induce in me an uncontrolled disposition, a mood, like that induced by rainy or sunny weather or by the phases of the moon, or by some mind-twisting drug. I might adore this "discovery," this "fact"; or I might detest it or neglect it.

It does not matter: my reaction would only have *psychological interest*, and so would any rationalization I could create through hindsight.

Thus I have demonstrated the imperious necessity of the Zetetican's "worlds."

On the other hand, how are they possible? Certainly I would not know where to start were I given the task of constructing a Zetetican-like "world," in the sense, say, of a precise and consistent global conceptual system. How could I even attempt to justify my choice of a point of departure without presupposing a framework to furnish me with the grounds for such a justification? To me, this initial but fundamental obstacle seems embarrassingly obvious and obstinate: In what I take to be the Zetetican sense of "world," which I strive to employ here, my "world" would necessarily embrace *everything*, comprising *inter alia* the only conceptual framework from which I can possibly draw the standards for evaluating anything. And this not only would include any appraisal of the validity or veridicality of my total view, or the "reality" of my "world," but also the simplest assessment of the meaningfulness or absurdity of this very question. At present I see no way for *me* to arrive at an *assessment* of my world without presupposing a frame of reference entailed *in* and *by* that world of mine. Were I to employ a *different* frame of reference (acquired from where?) and apply it to my world, in what sense of the phrase would it then be "my world" to which it was applied? Certainly *not* in the sense of my total view, my "system" or "synthesis," which is, after all, the traditional philosophical world for a consistent, all-embracing total view—with its logic, ontology, epistemology, value system, and so on.

I remember an American professor of psychology, a James Gibson, I believe, who generously supplied me with a magazine article wherein he had advanced some novel arguments for adopting *realism* as a total system: the real world was the realist one! I did not even have to look at the *content* of the article. How *could* there conceivably be anything recognizable as "reason (or arguments) for realism"? Either the argumentation takes place *within* a realist world employing a realist frame of reference, in which case the reasons or arguments may be reasons or arguments solely to someone already within a realist world; or the argumentation transcends the realist frame, in which case the validity of the argumentation obviously depends upon the acceptance of a *non-*, or *a-*, or *un-*, or *sur-*realistic world, with a corresponding conceptual frame of reference. Hence, if the explicit, cognitive integration of scientific knowledge is the most important dimension of perfecting knowledge, it is also the one marred by the most spectacular paradoxes. It seems that, as scientists as well as philosophers, we Tellurians must either choose a thoughtless, scatterbrained superficiality or, if we try to be thorough, complete, and consistent, end up in a soup a la Professor Gibson.

I regret that I must throw up my hands and admit that I never did graduate to a clear comprehension of why the Zeteticans did not fall prey to this devastating dilemma. But, obviously, *it would be a silly exercise in fatuous futility to attempt to prove the possibility as well as impossibility of the very idea of global conceptual schemes.* The validity of the result of such an exercise could be established only within one of an undetermined multitude of field dimensions and could be judged, at the most, to be tenable (or untenable) only by employing *some* explicit or implicit conceptual framework. That framework, when consistently perfected, was bound to tend toward totality, to become a philosophical system, a "world."

I have gone over all this rather thoroughly in order, among other things, to make it clear how difficult it would be to determine what the Zeteticans got out of me. Obviously what they would "get" would vary in correlation with the "world" within which all that was absorbed, which I naively would consider simply "pieces of information." I can only try to give a brief account of what I, to the best of my recollection, fed into the great "thinking machine" in Zetetica's interior. Only on what I dispatched can I report, not on what the Zeteticans received.

September 7, 1975
MYSTERIES OF MAN: THE CONSCIOUS CADDIS FLY

My dear reader, if you have had the patience to accompany me this far, I imagine that you shall find it much less of a strain to follow me the rest of the way. From here on my notes are bound to be less ponderous and perspicacious, for several reasons. One, about which the Zeteticans warned me, is that a rapid psychophysiological deterioration set in about thirty years after my return to earth and has been accelerating since my 178th birthday! I had noticed it already at home, in Hilary Hall. And so did my wise and gracious Italian wife, Alicia, and possibly even my children. That is why I left. But more about this later.

The second reason is that the topic is of more general human interest: How is it possible for any conscious, intelligent being to live and breathe his allotted number of breaths facing the stark reality of the inevitable, incontestible certitude of being no more?

Third, the Zeteticans seemed to make a particular effort—with the aid, I assume, of their enormous popularization machine—to lower themselves to my level and to adapt their verbal interchange to what they took to be my implicit "world," my conceptual framework.

But what could I tell them?

I started by talking about religion. I tried to convey to them the startling fact that otherwise sane and mentally competent humans would flatly deny the origin of Homo sapiens as a haphazard contingency, a fortuitous accident in the randomly risky raffle of survival on evolution's enormous wheel of chance and instead seek refuge in myths about some all-mighty, all-wise, all-benevolent being (or beings) who created the world —and man with it—for some purposes or other, unknown to all but that being itself.

No one, I assured them, in either my father's or mother's family was in any way prone to succumb to such superstitions. But some of the servants at Hilary Hall were. They went to church outrageously early on Sunday mornings, and I once went with them. I was madly in love at that time with a young chambermaid named Mona. She was from the Shetland Islands and was frightfully Christian. I asked her why she prayed in church. She said she prayed for my safe voyage to Pernambuco, where I was to join my father in a few weeks to inspect a coffee plantation he had bought in Brazil. "But," I said, trying to detect some order in the madness, "surely that must be blasphemy. To pray is to imply that this God of yours actually goofed when he created the world and, like some lousy plumber, has to come back to fix things up again—maybe, with a little miracle or something." She almost cried. She was beautiful. Now, she explained, not only would I perish on the sea, I would most certainly go directly to Hell and suffer eternal torture. I felt sorry for her and said no more. But I was glad I was leaving shortly. After she had so glaringly exposed her muddled mind, I no longer had much of an eye for the beauty of her body.

The Zeteticans quietly interrupted my retrospections. They were not, they said, interested in our servants' primitive superstitions. They were concerned with humans, they said, who were not illiterate, or cowardly, or intellectually dishonest.

Here, it seemed to me that the Zeteticans were naive. They made no allowance for the individual human's aptitude for distraction; laxness of determination; lack of precision and consistency; indefiniteness with regard to beliefs, thoughts, and feelings; and so on. Religious myths should never be seen as a clear and deliberate choice of a comforting life-lie, with a possible exception made for Blaise Pascal. To most, it is mainly a reservoir of unrelated verbal responses that serve as substitutes for rational arguments in mankind's bewildered demand for meaning, order, and justice—in "the world," and in "life." Why such a demand should exist in human thought at all is a mystery to me. Obviously *meaning, order,* and *justice* are words that are conceptually limited to certain aspects of some human enterprise. It would be what recent philosophers refer to as a "category mistake" to try to apply them to "the totality of human existence as such"—let alone to "the world"!

The Zeteticans, however, did not tend to share my feeling of puzzlement. It would seem quite natural, they suggested, for any individual conscious caddis fly to try to compensate for its absurdly brief existence by attempting to see its life in a means-to-an-end relationship within some wider context. That is, as the Zeteticans saw it, exactly what individual humans were, and still are, attempting.

I was, in a way, compelled to consent. But I also made it clear that, although their point was well taken, it could scarcely be construed as a *justification* for any metaphysical "meaning-of-life-and-world" demand. They had merely, and at the most, made the enterprise psychologically intelligible, plausible, or perhaps even excusable against the background of human ephemerality. They agreed. Moreover, they explained, that sort of thing was just what they were after. To attempt to demonstrate the fatuous fallacy in the reasoning behind a metaphysical meaning-demand seemed to them nugaciously superfluous. Their concern here, I gathered, lay primarily with the ludicrously low level of ambition that, by and large, characterizes humans' metaphysical meaning-demands.

Some human beings, they believed, would consider their lives "meaningful" provided a certain psychophysiological state prevailed during most of the interlude between conception and consummation, namely, a state wherein (a) the parasympathetic nervous system was predominantly activated, rather than sympathicus; (b) acetylcholine is produced rather than novoadrenaline (from the ganglions) or adrenaline (from the suprarenals); and (c) the blood is being rushed to the intestines and genitals rather than to the brain and muscles. "Of course," I replied in astonishment, "you have described a state to which humans would generally refer as pleasure, happiness, joy, bliss, contentment, satisfaction, cheerfulness, comfort, ease, peace of mind, serenity, tranquillity, and (more recently) state of thriving. Almost all humans, but particularly Englishmen, see these sorts of things as the principal and prime justification for their existence. *They make life worth living.*"

There was a long pause before the Zeteticans responded. For once, I thought to myself triumphantly, *I* have stumped *them*! Even with these innumerable conceptual schemes at their hands, they were obviously hard put to find a single one within which to make any sense of such ravingly eccentric value priorities. They finally concluded that I had exaggerated and oversimplified—which I had. Very few humans, if any, keep a clear-cut hedonistic budget, listing the length and intensity of positive and negative experiences, so that, when lying on their deathbed, they can figure out the degree to which they may be justified in assigning a meaning to their lives. First of all, many of them will not restrict themselves to "happiness" and the like. They find a stable acetylcholine state tedious, boring, and even unworthy of men. They require a somewhat adrenaline-spiced existence. And seek their sources of excitement and thrills in a miscellaneous menagerie of diversions and pastimes: mountain climbing, glider flying, rodeo, horror movies, merry-go-rounds, roller coasters, earthquakes, bingo, bowling, baseball, big-game hunting, or golf; scandals, distant disasters, contract bridge, car racing, crossword puzzles, curling, professional roller-skating, hockey, LSD, television, meditation, eroticism, free love, narcotics, sexual orgasm, promiscuity, pornography, perversity, obscenity, bigamy, monogamy, commercials, protest marches, beauty contests, music festivals, charity bazaars, weddings, funerals, and other processions and parades, teach-ins, love-ins, be-ins, grand masses, tenure denials and executions, pot smoking, glue sniffing, suicide and other "kicks"; terrorisms, hijackings, wars, conspiracies, rebellions, revolutions, church work, fireworks, and ladies' teas (with hats). There are even those who titillate themselves by attempting to achieve a total awareness of the actual human condition and then exploiting their insights into the absurdity of life as means to "heighten the sense of living."

But as I repeatedly emphasized in my report to the Zeteticans: It isn't *that* simple. Once more I drew their attention to the peculiarly underestimated cognitive as well as empathetic disintegratedness of the human consciousness. The Zeteticans, however, appeared strangely reluctant to accept fully my claim that we humans—I have found this to be the rule rather than the exception—are admiringly capable of simultaneously maintaining a multitude of logically incompatible positions. And since the whole field is so thoroughly muddled, it is with the greatest ease that the average human can permit himself to remain totally untouched by an argument that, by more objective standards, ought to be devastating to his chosen priorities. He may even distort such a counterargument so as to make it produce the *psychological* effect of a decisive *pro* argument for a *favored* position. This is, I suppose, a mechanism that is not entirely unrelated to what some recent psychologists would call "cognitive dissonance." Or as my old teacher Arthur Schopenhauer would have put it: *"Wer ueberzeugt wird wider Willen, bleibt seiner Meinung doch im Stillen"*—"The empathetic disintegratedness may or may not be independent of its cognitive counterpart." There could well be a form of feedback here. But, essentially, empathetic disintegration is different. It is the rather mysterious psychological mechanism that effectively prevents a piece of lethally destructive information from penetrating into a person's volitional layers, where it could cause irreparable damage. It is permitted instead to remain comfortably on the intellectual surface.

When it comes to retaining a hedonistically acceptable state, with a parasympathic acetylcholine predominans, there cannot be much doubt, I assured the Zeteticans, that the most effective metaphysical ontological hebetant is mortal man's knack for extracting brief intervals from the total term of his existence and filling them either with pastimes like the rather special ones listed above or, more often, with the safest, greatest, greyest, and most generally optunding inappetence—labor, that is, "useful" physical or mental exertion, often with an added ingredient of superficial, nonintegrated belief in myths about God and an afterlife. Either one may be used as opium for the People. "Work," reads the constitution of the USSR, "is the duty of every citizen, according to the principle: He who does not work, neither shall he eat." "Produce!" cried Carlyle: "Were it but the pitifulest, infinitesimal fraction of a product, produce it in God's name." There are already areas on earth where 2 percent of the population will soon be able to produce more than the other 98 percent can possibly consume. Yet Tellurians shall have to kill 600 milliard more free hours. Leisure counsellors will have more than petty weekend neuroses on their hands. Aristotle said that a society, unprepared for true leisure, will degenerate during good times. Too much leisure with too much money has been the dread of societies throughout the ages. That is why nations cave in. The German futurist Robert Jungk devoted a series of television programs to a study of work addicts and their "leisure-osis" in more advanced societies where *"die Zukunft hat schon begonnen."* His solution was to form quartets, repair or build one's own television set, or better still, build *completely useless machines*—something like Alexander Calder's mobiles with built-in motors. Whether this may be a better answer than anything that labor-plus-myths can offer seems to me a toss-up. However, it doesn't take a psychologist to predict that if humans try to fill their leisure time by, say, putting a small, white ball into a slightly larger hole, they will, as a people, go quietly nuts. The time is close when professional golf, baseball, football, hockey, wrestling, and roller derbies just won't do to keep the labor force under a sufficiently permanent sedation. An increasing number are seeking higher thrills—or more thrilling "highs."

> Someone said he had a friend who liked to shoot model airplane glue. No one else had heard of that. Sniffing glue, yes; but not shooting it. They had heard of people doing something to paregoric and shoe polish and then shooting it, but the high was reported to be no good. Heroin, of course, was the best. Heroin and a *bombita*. It gave the best high, completely relaxed, not a problem in the world.
> "But that's not really the best high," one addict said. Do you know what the best high *really* is?" The voice was serious. Everyone turned and stayed very quiet to hear, maybe, of a new kind of high that was better than heroin, better than anything else. "The best high"—the voice was low and somber—"is death." Silence. "Man, that's outta sight, that's somethin' else. Yeah, no feelin' at all." Everyone agreed. The best high of all was death.*

The only snag in this whole line of reasoning is its pathetic oversight of a most disheartening fact: that there will be no one in that defunct cadaver to enjoy this "best of all highs."

Even more obvious, of course, is the impotence of religious myths. Atheism is extinct in the more advanced parts of the world—for lack of opposites. Even in my student days a serious atheist was considered a slightly ludicrous bore. Religious myths were neither pompously condemned nor solemnly repudiated but received as charming subjects for art and poetry—like old-fashioned steam engines are today, or antique hot-water bottles. Needless to say, the myths have, in this form, lost all potential for consolation.

On the other hand, an excessive stress on logico-rational knowledge

---

*This passage is cited from an old *Life* magazine that was lying about. The aristocratic scholar had probably found it beneath his academic dignity to enter such a plebeian reference in his notes.

perfection may in itself serve as an effective means to hebetate the death awareness in man and prevent the growth of the courage for dread and anguish. A case *ad rem* is Lev Tolstoy's *The Death of Ivan Ilyich*, a most impressive demonstration of the fact that the only decisive, crucial criterion for true insight into one's own fate is *the internalization of the awareness* and *not* tenable evidence or crystalline clarity. Ivan was thoroughly convinced of the logic of the syllogism: Caius is a man; all men are mortal; therefore Caius is mortal.

> That Caius—man in the abstract—was mortal, was perfectly correct, but he was not Caius, not an abstract man, but a creature quite, quite separate from all others. He had been little Vanya, with a mamma and a papa, with Mitya and Volodya, with the toys, a coachman and a nurse, afterwards with Katenka and with all the joys, griefs, and delights of childhood, boyhood, and youth. What did Caius know of the smell of that striped leather ball Vanya had been so fond of? Had Caius kissed his mother's hand like that, and did the silk of her dress rustle so for Caius? Had he rioted like that at school when the pastry was bad? Had Caius been in love like that? Could Caius preside at a session as he did? Caius really was mortal, and it was right for him to die; but for me, little Vanya, Ivan Ilyich, with all my thoughts and emotions, it's altogether a different matter. It cannot be that I ought to die. That would be too terrible.
>
> If I had to die like Caius I should have known it was so. An inner voice would have told me so, but there was nothing of the sort in me and I and all my friends felt that our case was quite different from that of Caius. . . . And now here it is! he said to himself. It can't be. It's impossible! But here it is.

Finally the truth filtered through to Ivan, and suddenly he realized in his bones and marrow that his malady was not merely a matter of a diseased kidney but of leaving behind him as pointless a life as any other life and facing ultimate and total obliteration. "*For the last three days he screamed incessantly.*"

"Nothing," said my fellow student Sören Kierkegaard, "is to me more ludicrous than a man who rushes to his job and rushes to his food; when a fly rests on the busy executive's nose, or the draw-bridge goes up, or a tile falls down and kills him, my heart rejoices with laughter. He reminds me of the old hag, who, when her house was on fire, only saved the poker. What more does he save out of his life's conflagration?"

Ivan Ilyich was exactly such an assiduous executive. The major "welfare" problem facing the conscious caddis fly, Homo ephemeralis, is the fact that the enormous expansion of mass education plus the explosionlike increase of leisure time will permit present and, not the least, future generations to anticipate Ivan Ilyich's shriek with a margin of perhaps thirty to sixty years, which will make for a relatively long shriek. The question is whether it is morally right or psychologically possible to forearm man against such pernicious insights, and whether it is in accordance with human dignity to permit, or even tempt, man into some sort of more or less attractive, sophisticated, high-brow life-lie at the expense of his full humanization. This, I said to the Zeteticans, is *one* aspect of the predicament inherent in a conscious caddis-fly existence, and one that few humans have any inclination to face.

September 11, 1975
MORE ON THE MYSTERY OF MORTAL MAN

Reluctantly anticipating my physical, as well as mental, enfeeblement, I have carried as many of my possessions as I could possibly manage up to the room I have somewhat arbitrarily chosen to be my bedroom. I have taken, for instance, half a barrel of slightly yeasty blueberries, dried rowanberries, and rose-hips. I even made an improvised desk of the unhitched kitchen-scullery door, which, with no little effort, I contrived to place over the top part of two of my bedroom's chests of drawers. Happily, all of the rooms have running water, so that is no problem. However, I am bound to confess that I still do not quite realize in my bones that I am here within these narrow walls . . . to die. This is partly because I am probably as skillful as the next man in debarring such a devastating comprehension from penetrating my volitional layers. Also, as I have mentioned before, I am still hoping against hope for the riders on the white horses—my Zetetican saviors. Finally, in my silly immaturity, I still desperately try to find some comfort or consolation in the fact—which I hope I shall be in the position of relating to you some other day, if I live long enough—that I have an impudently healthy duplicate running around on Zetetica. (Busily rewriting the whole scenario of the universe, no doubt. Or something else—I don't know.)

Neither am I sure why I went on and on the other day about this Ivan Ilyich. I never knew Lev Tolstoy. He belonged to the generation after mine. I have only read him since my Tellurian return. Sickening stuff, I find—most of it—a pollyannaish *feuilletoniste*. He seems to have been a syrupy sentimentalist who, as if he were English or something, believed that if only acetylcholine (happiness) could prevail over novoadrenaline (anguish, pain) during most or all of the (terminal) existence of most or all (terminal) beings, then we would have reached the ultimate end and goal of all conceivable human concern—the *summum bonum*. Not that I have anything against people's being "happy" or enjoying themselves—if they can. After all, we are on earth and there is no cure for that. There is nothing gained by peevish whimpers or by mustering a monstrous moral outrage against the universe's breezy, nonchalant, insouciance regarding human suffering. To pompously "accuse" the universe—this myriadic whimwham menagerie, this geophysic-astronomic childrens' knickknack mechanical I-don't-know toy—of injustice, virulence, cruelty, shamefulness, malignancy, or absurdity is to bestow upon it an undeserved compliment of humanness. The universe isn't even that.

And were any reflecting consciousness, motivated by merciless truth-seeking and by the opportunity for unrestricted insight and knowledge perfection, inexorably driven to face the fate of mankind in this cosmic carrousel, in this rapacious roller coaster—one should at least expect a slight drop in the number of enthusiastic joiners. But *not* because the human bodies are lacking in acetylcholine, happiness, peace of mind, or whatever. That is not the point. I think my friends will testify that I am perfectly capable of all sorts of merriments; I enjoy research, as well as less intellectual pastimes. ("What can a man do," asks Hamlet, "but be merry?") The point is, rather, that in retrospect, or seen against the backdrop of the ephemerality and terminality of human existence, to relish the so-called positive qualities of life, of which so many good Englishmen seem to think so highly, makes as much sense to me, after my Zetetican sojourn, as does the condemned prisoner who relishes the prospect of selecting the most tantalizing and scrumptious dainties for his final gourmet meal. *Sub specie cadaveris* such subtleties escape me. Could anyone seriously be concerned with the degree to which what is now a corpse previously enjoyed or did not enjoy whatever at present rots in his belly? By the same token, I frankly could not care less whether any beings—a caddis fly, a sea turtle, or an instance of Homo sapiens ephemeralis—is judged as enjoying a predominately comfortable, acetylcholine-filled interlude from time t to time t', i.e., the so-called life-span of the species in question.

So, why this long harangue about Ivan Ilyich? I am not sure. Perhaps it is simply because that light paperback was the only book I had brought with me. There were quite a few other books I considered—*The Meaning of Death*(!), *About Death and Dying, Answers to Questions about Death and Dying*—but they were all so asinine, exposing such an embarrassingly blind and naked ignorance, that there was little choice. The authors of those books did not have the faintest flicker of comprehension of death, the being-no-more. Like religions, what they actually offered amounted to nothing beyond infantile philosophy: How to "accept dying" without the Ivan Ilyich shriek and "go gentle into that good night." But Ivan was right of course: If you understand what it means to die, a shriek is the only psychologically adequate reaction, evincing an awareness in one's bones of death and its all-obliterating consequences. In the words, as I remember them, of a brilliant young Welshman* whose poetry my children recite by heart: "Old age should burn and rave at close of day;/Rage, rage against the dying of the light."

There is one respect in which the case of Ivan Ilyich is singularly exceptional: Most humans manage only too well to guard themselves against such grim and gruesome cadaverous insights as could vivify an Ilyich shriek. Whenever humans die, it is almost invariably under the accommodating carpet of confusion and perplexity, due to some more or less rapid psychophysiological decomposition, plus pain-relieving myths and medications. And should one be unfortunate enough to survive for another round or two of the heart and even prolong the allotted life-span beyond normal expectations, there is always the merciful, magnanimous savior: the benign silliness of a blissful senility. Their faces glowing with a good and serene conscience, all the grievers and mourners can now sing hallelujahs at the transmutation of the well-nigh corpse into a flattering memory. And the cadaver, it seems, silently approves: "It was, after all, the best that could happen. It was best this way."

In other respects, however, Ivan Ilyich is indeed typical. He was like that busy executive whom Sören Kierkegaard ridiculed. He even presided over meetings. He was an important man, *one with duties*. The duties cut his life into short time periods, each with a gravely serious goal to be reached. The goals in turn assigned meaning and purpose in abundance to any enterprise undertaken within each goal-determined portion of his life. And would it not be tempting for Ivan to argue that since his life consists of a limited set of fragments of definite duration, and since each fragment is filled with meaning and purpose, then the same must go for the total set—his whole life? And despite the fact that Tolstoy does not in the end

---

*The "brilliant young Welshman" is, needless to say, Dylan Thomas; and the poem is one of his best known: "Do Not Go Gentle into That Good Night."

permit Ivan to fall into this trap, the impression is inescapable throughout: Ivan never questions the *purpose* of his existence as long as his health permits him to engulf himself in what he sees as his mission. He has found his niche. He keeps to his part, parading his path of importance, posing in the chair, devoutly playing his role, as he sees it, in strict accordance with social norms and expectations.

Here, I told the Zeteticans, we have in Ivan Ilyich exemplified what I take to be the most fundamental mechanisms to allow a life-long delusion of sense, significance, meaning, purpose, aim, design, etc., to permeate the human caddis-fly existence. That in turn confers importance, earnestness, and solemnity upon it and renders it so ponderous that we are forced "to be serious about such mean allowance of breath as life is" and, furthermore, to "waste the evening commiserating with each other about the unhygienic conditions in our worm-cases."

In point of fact, when faced with the profoundest primordial trauma —the human cosmic-ontic condition—and seeking refuge in the great universal repression that bars and bolts all fatal insights, the most popular *missions* into which humans plunge themselves with the ecstatic passion of a panic-stricken 'possum are indeed generally concerned just with attempts at improving ecological, social, economical, political, medical, psychological, sexual, matrimonial aspects of the "conditions in our worm-cases." Like oversensitive children who cannot bear to look upon a dead bird, humans draw their missions up before their eyes like a blind to shelter themselves from disturbing insights into their own fate. And it invariably works—for a time, for even a lifetime—so long as the attempts are unsuccessful. But if the missionaries' utopia were ever to be reached and famine, starvation, corruption, and oppression were all gone, so would the blissfully sheltering blinds also be gone. This cosmic-ontic condition of humanity is traditionally neither perceived nor heeded by other than members of what once were called the "upper-classes," who had a higher education and an abundance of leisure time. The recent desperate need among psychologists and psychotherapists for a "philosophy of man and his fate" arises exactly from the general improvement of living conditions and education. Fifty years ago there was no problem for psychotherapists about what would constitute a cure or, at least, a step in a more "healthy" direction. Freud's patients were largely suffering from heavy hysteria, dramatic paralyses, or an inability to talk or move. The more advanced countries today have caught up with many utopian ideals concerning poverty, political oppression, and psychopath-creating authoritarian family structures, while at the same time belief in gods and devils, heaven and hell, angels and immortality have waned. In these countries people suffer less from nightmarish misery than from the more subtle disorders previously buried by the harsh and bitter struggle for existence. The clinical psychologists are unexpectedly confronted with patients who, by all socio-economic criteria, are tremendously successful and well adjusted. They have merely—prematurely, as it were—anticipated the dying groan, the Ivan Ilyich shriek: What is it all about? Thus what once was an obviously commendatory endeavor—to abolish poverty, ignorance, and injustice—is gradually presenting for the human caddis fly a problem the severity of which will increase in correlation with the growth of leisure time and socio-economic and educational progress. That problem is embodied in the most humanly relevant questions—if indeed these are questions—of all: *What does it mean to be human? What is the lot of mankind in the cosmos?*

September 17 (Justus' birthday!)
MEETING MYSELF

I have been bedridden for a couple of days. Nothing serious. Merely dying. The little of me, that is, that there is left to die. I saw myself in the Hepple-white-imitation looking-glass in the downstairs sitting room. I looked like something from a heap of corpses in a Second World War extermination camp. So I crawled back to bed again, contemplating what to write next, if I were ever again to acquire the energy. Today, since I have dined on an assortment of berries, with tap water for wine, I feel almost up to it.

Incidentally, I have been troubled now and then by a nagging suspicion that this obsession of mine with writing down my story has become *my* mission, *my* blinder against perceiving the absurdity of human existence. Why, it is not even absurd! It just *is*. And too: it is *not*. That is all. Now: *How could I possibly be blissfully blessed by a blind that shows nothing but an honest and faithful picture of exactly what I most dread to see?*

But why, then, did I return to such a wild, banal, grotesque, and loathsome carnival in the Tellurian graveyard, as human life truly is when seen *sub specie cadaveris?* Was I not somehow hoodwinked or blinded?

I have, I believe, explained one of my grounds for leaving Zetetica.

But I shall return to that later. What I have neglected to mention is the unflattering fact that I failed. You understand, the Zeteticans were not merely satisfied to have collected a living fossil. I qualified splendidly for that. But they had expected to find a fossil that would still be open, as it were, to evolutionary change. I was too fossilized; I did not change in ways recognized by the Zeteticans as steps toward making possible improved communication. Seen from the Zetetican side, our exchange of ideas must have looked analogous to the aforementioned zoologist attempting a pleasant chat with his pet porpoise. So what did the Zeteticans do? They simply grew another me. They took cells from my body, bathed them in what I take to be some liquid nutrient and little by little grew a replica of my body. It must be, I think, a similar process to that now well known on earth as "cloning," except that all we have cloned here, as far as I know, are carrots. Although I have recently been told that there are experiments going on with animal cells. Anyway, on Zetetica there were clearly no problems in producing human clones. The master trick, they proudly brayed, was to reproduce and imprint a facsimile impression upon the clone's virgin brain that would perfectly represent the total amount of all my memories and skills. Again, the necessary process was transacted and consummated through those confounded thinking machines, or whatever they were. The whole thing was quite tedious and time-consuming. But Zeteticans were more than amply supplied with time. And with regard to tedium, the concept—at least as I know it—seemed nonexistent on Zetetica. It took several hours daily for more than a year, with my head in something like a modern hair-dryer, before I was finally introduced to—myself. It was a full-grown but peculiarly young-looking "I." Not me as a teenager, though, or in my twenties, but at about the age of fifty, with wrinkles, eyebags, and scars removed. I liked my looks!

Now, the Zeteticans' idea was for me to *identify* with my clone. It ought to have been easy. We were already identical. *My* self-identity was *his* self-identity. The old body would be put out of circulation. The whole change would appear to me as though I had just awakened and found myself with a brand new body. And as my present skills and memories should occupy scarcely more than a fraction of the cloned brain, there would be a better chance for me in time to reach a near-Zetetican level of comprehension and thus to mature beyond my present, primitive humanness. Thus I would be able to serve not merely as a source of paleontological information but as an active participant in a Zetetican enterprise. The

temporarily rejected old body would not be permitted to perish. The Zeteticans were no squandering wasters. It would be kept indefinitely in a condition somewhat like suspended animation, I was led to believe. *"Just in
case,"* I suppose they would say.

September 18
RETURNING TO ROBIN HOOD
AND THE SHERIFF OF NOTTINGHAM

It is getting cold. I must have the heat on most of the time. Only when the
sun shines do I turn it off. With the sun bathing my face I can go to sleep
here in my wooden chair, uncomfortable as it is, my neck resting on the
top of the back. I awaken, shuddering, chilled to my bones, and aware of
how readily I might have gone to shivers, then and there, having taken my
last sleep. My brain's oxygen intake is still sufficient to prevent me from
greeting my imminent death with euphoria. I do not want to cease to be. I
am such a creature of habit. I am so used to being alive. I find living a
particularly habit-forming activity. So, I suppose, deep down somewhere I
am waiting for Zeteticans to come. But I'm kidding myself. They thought
Arthur Schopenhauer was too far gone. Who on earth or Zetetica could
take any interest in this chair full of old bones and rags?

How *could* I—I asked myself over and over again—how could I have
been so stupid as to return?

To put it bluntly, I had simply declined the invitation to "mature" any
further. I refused to abandon the old "myself" in order to make another
leap when the time was considered ripe to a more adult, sophisticated
"myself." The Zeteticans accepted my request to leave, with conde-
scending sympathy and compassionate forbearance. I must have seemed
as ludicrous to them as any adult who would not give up playing Robin
Hood and the Sheriff of Nottingham. But I was bored. That was the
difference between my clone and me. He was eager to "mature." True
enough, we both wanted to unmuddle some of the multitude of puzzle-
ments that seemed suggested by Mitya's* mysterious phrase, "What is it
all about?" But it caught *me* deep in the grips of a nostalgia unshared by
my clone. He realized the obvious: If there are in the world any questions
worthy of answer, then the circumstances under which such questions are

---

*The oldest of Dostoyevsky's Brothers Karamazov.

likely to be effectively rephrased and answers provided would be like those prevailing on Zetetica: (1) an indefinite number of conceptual frameworks or "worlds" within which the questions may be cast; (2) an indefinite tenure of time within which these sophisticated consciousnesses and their elaborate thinking machines could suggest and sort out the answers to such questions—if in the end they turn out to be questions.

The preservation and perpetuation of a ratiocinating consciousness were, needless to say, a *sine qua non* for any Zetetican enterprise, whether philosophically significant or not. But I had found myself thinking that the Zeteticans were more than naturally and necessarily concerned with survival. The nearest *possible* threat to their existence was at least nine billion* years away. That was a bit too far in the future to hold my undivided enthusiasm. I still had in me the urgency feelings of *ephemerae*. To rephrase Ivan's† famous dictum: I did not want millions of years but answer to my questions—as silly as that sounds.

And there were, I am bound to confess, other things I wanted too.

"Robin Hood" was not the only game unknown to Zeteticans. They did not even "play house." There was no sex, of course. Not that it was impossible. There were still sex differences on Zetetica. If the necessity should occur, they were still prepared to regress to propagation and try to preserve their species through generation shifts. In other words, in the unlikely event that their present form of existence should, for some unimagined reason, be obstructed or blocked, they did not rule out the grim contingency of returning to *samsara*—the revolting, revolving, unmerry merry-go-round of birth and life and death and birth and life and death.

At present, however, they were no more excited about "playing sex" than "playing house" or "Robin Hood and the Sheriff of Nottingham." There was, as I said, minimal sex difference. But no Zetetican felt any inclination, it seemed, to shout an exalted *"Vive la difference!"*

So, no sex. And I, for one, was far from sharing my master Dr. Schopenhauer's contempt for women. On the contrary, I spent long hours in the Zetetican "hair-dryer" wallowing in reminiscences about romantic adventures all around the sex-loving planet of Tellus.

I was infantile, a ludicrous anachronism, maybe—but a very home-

---

*These are real billions (not milliards), i.e., a million million.

†Ivan, the second of Dostoyevsky's Brothers Karamazov.

sick anachronism. "You *promised* me," I reminded them. They acknowledged that. I would be returned to earth, if that was what I really wanted. They just could not understand me: "You will die! If you leave Zetetica to take up the caddis-fly life of Tellurians you will have about thirty years. Then rapid deterioration will set in. Here you can live indefinitely."

I felt like that pompous pumpkin-head who cock-a-doodled "Give me liberty or give me death" when I, satiated with silly pride and sentimentality, said to the Zeteticans: "You have my pure, taintless, undefiled and uncorrupted replica. *He* can have your eternity. Give *me* thirty years on earth!"

Big words. Only now can I appreciate the inapprehensible, moronic, fatal folly in what I then took to be merely a rash but heroic decision.

I was going to say that I have repented the decision ever since. But that is not entirely true. There were times when I, like other humans, forgot about my fate and lived a moment as if it were an eternity. Some might be tempted to ask me: Was it worth it? But that is the wrong question. Shortly a corpse will be occupying this chair, and there will no longer be an "I" to find it "worth it" to have traded thirty earth years for an almost certain immortality on Zetetica. But as for the rest of my life, mark my words: this chair shall be my mourning bench of the most bitter repentance. Make no mistake about that!

On the other hand, as I said, there were moments.

Thanks to my thoughtful Zetetican friends, after some intricate manipulations with which I shall not bore you, I was economically independent. Equipped with a German passport and a certificate stating my birth as August 3, 1916, in Krümmel, I took to "the world" like a starved lion to a piece of raw filet mignon. I traveled, mostly in Europe, but also in Asia, the Pacific Islands, and the Americas. It was in Victoria, British Columbia, that I happened to get my hands on a local Swansea paper, according to which "the second oldest building on Anglesey," Hilary Hall, was for sale. I bought the building. Then I redecorated it and refurnished it in carefully selected, light, modern Scandinavian (except for the genuine Louis XVI dining room). It got a new coat of white paint, real gold, and red silk upholstery. It brought back childhood memories.

However, I shall not slide into sentimentalities but will endeavor to make the rest of my story brief, as it should be. It really is of no significance. I only thought you might like to know how things turned out. Moreover, it is easy writing; it does not strain my brain.

September 27
MY SECOND COMING

I tried to crawl downstairs about a week ago. I seemed to remember having seen a glass of geriatric vitamin pills in the kitchen—not that they could do me much good, but they could hardly hurt—so I tried getting them. And, of course, I fell down the stairs. I was knocked unconscious; and when I came to, I was nauseous and had a terrific headache. Miraculously, nothing seemed broken. I found the vitamins. Now my menu was geriatric vitamin pills, water, and, after I managed to crawl upstairs again, my assorted berries.

I have been reclining in my bed most of the time since I recovered from the fall, pondering my second lifetime on earth. I suppose that if a life could be properly said to have been happy—which, as I have shown, is a notational impossibility—I am sure my second one would qualify better than most. I really went all out playing Robin Hood and the Sheriff of Nottingham! In my wild, inconsiderate, infantile recklessness, I married and made my microscopic contributions to no fewer than four human caddis flies.

I originally grew up during a peculiarly pronounced romantic-rationistic period. Incomparably the greatest hero of my youth was Johann Wolfgang von Goethe. He was a distant relative on my father's mother's side, which gave my father a pretext for visiting the genius and introducing me to him. I was twelve at the time. Goethe lived in Weimar, where he was in charge of the ducal theater. It was exactly a year since he had finished the first part of *Faust*. I was completely entranced and was soon thereafter to enter a rather late period of *Sturm und Drang*, dressing like Werther* and looking for some Lotte† with whom I could fall desperately and unhappily in love. But always in vain! By sheer coincidence—I could scarcely have been that irresistible—my ever-so-desperate love was invariably returned. Nevertheless I somehow managed, following my master Dr. Scho-

---

*Von Habermann is, of course, talking about the main character of *Die Leiden des jungen Werthers* ("The Sorrows of Young Werther"), which was very popular with upper-class European youth of the late 18th century, the height of the *Sturm und Drang* period.

†The lady, Charlotte Boff (or Wetzler), for whom Werther kills himself.

penhauer's advice, to stay unmarried during my first Tellurian residence. And on Zetetica they were not "playing house" at all!

What with one thing or another, I seemed to have retained a good deal of my early romanticism vis-à-vis women. I was inclined to adore them at a distance. For not at any price would I run the risk of being classified now as a "dirty old man." So in spite of my natural extroversion— people generally find me rather entertaining—I must have appeared to young women as reticent and reserved to a degree.

I was both envious of and encouraged by a Danish friend, who was six years older (counting my own age from 1916) and who seemed to have no inhibition against getting involved with young women about the age of his own children. Even so, I am bound to admit that I was struck by the ludicrousness of his behavior, which I found fundamentally indistinguishable from that of a rutting dog running around with his tongue hanging out and scenting a neighborhood bitch in heat. His grey beard and his title of "University Professor" added only slightly to the drollery of his derisory burlesque.

And yet, if I were in all honesty to look at my own attitude through the same Zetetican spectacles, it might well be a rather troublesome task to resolve any questions as to which one of us were to be more justly considered the champion laughingstock. My desires were, from a Zetetican viewpoint, equally primitive, based on parasympathicus/sympathicus or acetylcholine/novoadrenaline priorities. Mine were merely cluttered up by this peculiar eighteenth-century *Sturm und Drang* yearning for a sublime and illustrious amorous tragedy to *ennoble* my caddis-fly existence, enabling it to rise above trivia into something exquisite, extraordinary— Dante and Beatrice, Romeo and Juliet, Werther and Lotte. Frankly, I think I might well be the one to run off with the grand silliness-prize!

Be that as it may, it so happened that both our asininities came to full bloom during a philosophy conference in Vienna in the autumn of 1946.

It had made me sad to see the alarming deterioration of philosophy since "my own time," as it were, and particularly since the Second World War. Some spectacular advances in logic and philosophy of mathematics notwithstanding, philosophy, conceived as the currently most common activity of members of philosophical institutions, was rapidly approaching its all-time lowest ebb—"linguistic philosophy," so-called probably because it is neither linguistic nor philosophy. It made me alternately sick and angry to listen to "philosophers" performing as though they were semantic sorcerers and soothsayers, as do-it-yourself greenhorn lexicographers, or as hopelessly outdated, but nonetheless dogmatic, linguistic dilettantes. And as if the piddling, paltry nugacity of their fatuous and philosophically futile enterprise were not enough, they had to display, in a tone of insufferable vaingloriousness, a propensity for perpetrating that peculiar, similie-profound, pseudoperspicacious clever small talk, which unhappily seemed to become the differentiating hallmark of almost all typical philosophical discourse in the years following what may have been sarcastically referred to as "the revolution in philosophy."

There were times at those meetings when I wondered if words had a meaning at all. Certainly they rarely were as *precise* as philosophers thought they should be, and often they were as meaningless as a bullfrog's chorus. Nine-tenths of all words were, I found, parrot noises; they were not weighed, savored, and tasted, but merely repeated. There were whole sentences that lay saliently on the surface of philosophers' memories and were tossed out ten thousand times a day—flipped off the tongue, rebounding from the tympani—without a thought to give them life. To nine-tenths of this vain nine-tenths no one listened, so they were doubly vain. A minute disturbance in the air! *Brek- Ek- Ek- Ek- Koax- Koax-.* Men talk, as I had explained to the Zeteticans, because men have the capacity for speech, just as monkeys have the capacity for swinging by their tails. For philosophers, as for other human caddis flies, talk passes the time away that would otherwise hang like a millstone about a man's neck. Tellurians in general, and philosophers in particular, swing from day to day by their long prehensile tongues, and are finally hurled headlong into their silent tombs or flaming furnaces.

One session had been especially trying. A dapper man in a morning coat, with a Spanish name and an Oxford accent, made the most of his "discovery" that "a cow that speaks flawless Greek is a logical contradiction." I finally lost my calm, raised my hand, and interrupted the speaker: "I don't quite follow you," I said. "*How* flawless must the Greek be in order to render the Greek-speaking cow a logical contradiction?" There was a long silence. "Surely," I continued, "if all that the cow could say in Greek was to pronounce one single Greek letter, say *mu—.*" The audience broke into laughter. I left.

It was a cold day. I went to the counter in the atrium of the old University and ordered hot lemon tea with cognac. I was joined by my Danish friend, who introduced me to *his* latest "discovery," a lovely Italian girl in her twenties with the sonorous name Alicia Salcia Arrigo. I took one glance at her and the whole primitive generation-shift machinery immediately activated. "Love at first sight," or whatever the silly thing is called, went all over me like nettle rash. She complimented me on my "scintillating wit," and I gave further air to my disgust over the meetings, speaking a bit panicky, maybe, in the hope that my "rash" would not show.

Had Werther finally found his Lotte? Oh well, there was certainly no external resemblance. Alicia was a warm and wise woman, with copper luminant, brown hair; big, dark, "soulful" eyes, with an occasional roguish flash; a slightly arched nose with exquisitely aristocratic nostrils. Her sparkling smile melted everything within sight, and her laughter and gaiety were as contagious as the measles. No, admittedly, she did not fulfill the formula for a Lotte. Nevertheless, I assure you, for days and nights to come I wallowed in the most romantic and deliciously insufferable eighteenth-century sorrows: I had been set adrift on a raft of melancholy.

Until one day. My Danish friend was to leave for a more promising conference on Spinoza in Madrid. I had driven to Vienna in a Peugeot stationwagon that I had picked up a month before in Paris. And of course I volunteered to drive my friend and Alicia to the airport. She was seeing him off, so we saw him off together. And then we returned to Vienna.

I was hardly at ease alone with her in the car, and I entertained her tremulously with stories about our departed friend and our many joint escapades. Before I knew it we were in the middle of the tumultuous traffic on the Schubert Ring. Uncertain about the location of her hotel, I turned to her: "And now," I asked, "where can I take you?" She looked straight ahead. Then, with a microscopic indication of a smile, she replied, "What about the Côte d'Azur? It certainly is the season for it." I reeled behind the wheel. Had I heard correctly? What did she say? And if that was what she said, what did she mean? She smiled: "Oh, come, come, Hilary. You hate this conference, I know. So do I. I want to leave. My parents live in Switzerland. I don't want to go there; but I have a rather charming little house in the old part of Monaco Ville. You'll adore it, I'm sure. It's pleasant there now." So much for my eighteenth-century sorrows!

It was late afternoon before we got out of Vienna. I had hoped to make it into Yugoslavia before night, but found to my irritation that we would probably have to stay overnight in some ghastly tourist trap in the Austrian Alps—an immense, obscene, monstrosity of a "sports" hotel lit up the whole area. We looked around, hoping for something slightly less loud and flashy, and found a narrow road leading farther up into the mountains. But—motor vehicles were prohibited on it. Then we had a bit of luck. We discovered a bobsled run nearby—and no signs prohibiting motor vehicles. We drove on for almost two hours in pitch dark—and in first gear—before we finally found ourselves on the top of a mountain.

We did not expect to find much of anything else there and had resigned ourselves to make the best out of it, using our sleeping bags. Then we noticed lights: a house. We approached reluctantly. Could we knock on the door? It was two o'clock in the morning and we were on an otherwise deserted mountaintop. On the other hand, there was light and noise inside. We were still standing, hand in hand, in front of the door when it suddenly opened and disclosed an eight- or nine-year-old boy and a huge St. Bernard. Neither seemed to find our presence even a trifle puzzling. The boy smiled: "My mother is in the kitchen." And so she was, doing dishes. But she waved to us to continue into the next room. There we found ourselves confronted with four women with peroxided hair watching an old German movie on the television set, while a red-cheeked, healthy-looking young man ran around, obviously squiring the ladies and attending to their slightest whims. A table was pointed out for us to sit at. Then, without asking, the ruddy young man placed a ten-liter *Johannisbeere-Wein* container in front of us. We drank the sweet wine and asked if we could have something to eat as well. We were served a delicious mutton roast.

The television was still on when we stumbled to our upstairs bedroom. We were so dead tired that it was not until the next morning that we were struck by the senseless absurdity of yesternight's scene.

The sun was up. We hung out the windows, gazing over the precipice. Clouds floated below us. Here was a castle, there a tiny train, slowly winding its way. At the very bottom, a silvery serpent—the Danube. And for no reason at all, we laughed and laughed.

September 30
ENDING UP

Today, early in the morning, after having spent a straight fifty-four hours in bed, I climbed into my chair and sternly said to myself: "This is it. You know they are not coming just to collect this lousily upholstered structure of crumbly, shivery bones. Don't be silly!" But again I was baffled by the proficiency of this miraculous mechanism that constitutes the fundamental, but somewhat crepuscular, prerequisite for the possible existence of conscious cosmic caddis-flies like Tellurian man. Somehow I managed, in spite of all, to maintain a degree of doubt about that blatantly indubitable fact: *I am ending up*. How is it—I never cease to wonder—even *logically possible* for me to, as it were, "hide" such an unbearable truth from myself? I must know what "it" is that I hide. How could I otherwise hide it? If it now were possible for me to successfully "hide" "it," in what sense of "knowing what is hidden" can I be said to know what I have hidden? And if I am *not* successful. . . ? All I can do to throw light on the matter is to point again to the *indefiniteness* of what we intend to do, mean, and say by what we say. If humans were as well disciplined as the Zeteticans with regard to clarity, preciseness, consistency, and definiteness of intention, they would be deprived of any access to such temporarily life-preserving speciousness. What would happen then is difficult to predict. Maybe the amateur anthropologist is right who suggested that should man ever attain a state of total maturity—in sum, ever achieve the final, total, truthful disillusionment—then in all likelihood he would not keep going, but would simply lie down wherever he happened to be and with a long sigh return to the oblivion from which he came. I don't know. In any event, I am still sitting here. And if I live that long, I may even get around to explaining why I chose to leave my loving wife and children to sit here in solitude, moribund and macerated.

Had I but had it in me to be a bit more frivolous I might well have ended the last entry into my notes (of September 27) with a flippant, "Then we got married and lived happily ever after." I must admit that if that silly phrase were ever appropriate, then it would be to describe the twenty-eight years I shared with Alicia—and with our children as they came along, two boys and two girls.

But I was not in the mood for frivolities. Through my Zetetican spectacles I saw absolutely no moral justification for a human caddis fly to marry and multiply. On the other hand, those same spectacles seem to provide me with a sort of cynicism that relieves me of any guilt feelings. After all, only four more caddis flies to flicker through their allotment of minutes in a perplexed state of half-believed life-lies and confusion. No guilt feelings. Only *sadness*. In spite of its ludicrousness, futility, absurdity—*Good Lord, how I love them!* So there you are. How deeply and wholeheartedly I have redelved into Tellurian primitivity during these years—to the limit of allowing myself, without the slightest embarrassment, to enter into my notes such a banal, ignobly earthborn platitude.

It is even true: the original nettle rash proved incurable. On our third night in Monaco, I proposed to Alicia—if "proposed" is the word I want. I said, I think, something about the cosmic conditions of the human caddis fly, how dreadfully sickening I found it all, and that she could glitter all she wanted, I was not going to reconcile myself to our lot in the cosmos for 165 centimeters of wavering light. "Nevertheless," I concluded, "I shall be loath to forego one day of you for the next twenty-seven years or so."

She sat up: "Do I understand you correctly? You are suggesting that we spend the next twenty-seven years together?" I nodded. "Very well," she said, looking me over casually, "I accept. But why so exact, so specific? Do you suffer from some extraordinary, unheard-of disease or something?"

"Nothing unusual," I reassured her. "It's very common, but terminal. It's called 'life.'" And we dropped the subject.

We married June 12, 1947, in Monaco, but moved after our honeymoon to Hilary Hall, where we usually spent the spring and summer in the years to come.

The children came at intervals of three or four years: Taurus, Justus, Ina, Teresa. The boys are tall and dark—actually rather Spanish-looking. The girls are tall, slender, and Nordic. "A beautiful family!"—so everybody says. I had promised myself when I married Alicia that I should at least do my utmost to make it, by her standards, the best twenty-seven years for her any earthling could hope to sustain. And, needless to say, I had the same intention with regard to our children. There was never an easier promise to keep. In fact, the same *ought* to go for all my fellow caddis flies. I worked willingly with my warm and socially responsible wife in her various attempts to aid all those she took to be in any way "less fortunate," as the saying goes. However, there is no denying the fact that my temporary Zetetican residence had left me with a certain weariness of life, a black and gloomy despondency regarding all these well-meaning attempts to change the conditions in our worm-cases.

I cannot, *sub specie cadaveris*, wholeheartedly join movements to *reform* prisons, schools, the church, the society, or the alphabet. Nor, I am afraid, do I, with my clear intellect, fully grasp the dreadful significance of stopping communism, imperialism, Cosa Nostra, the Pentagon, oil spills, bad breath, sexual impotence, underarm wetness, indigestion, the growing of poppy seed, warts and hemorrhoids, the extinction of the blue whale, the destruction of the ozone shield by aerosol spray, or, for that matter, murder, rape, insanity, racism, terrorism, earthquakes, famines, slums, child battering, and starvation. Not that I could constrain my exultant hurrah the day we rid the earth of all this heinous infamy and malignancy. But so what? Would we not be merely depriving ourselves of some charitable divertissements well suited to accommodating our deepest desires to turn our eyes away from the inescapable and imperious fatality of the human condition?

On the other hand, how could we expect to do anything about the human condition unless we first overcome all the infantile problems connected with man's silly search for "the good life"—economic strength, social success, and political power? But also, of course, there is the matter of solving medical mysteries: Why do we die? And how do we prevent death? Then and not before can we shed our caddis-fly existence and take an active interest in what is going to happen to us; only then can we give up the Kantian, awe-stricken submission to a massive collection of hydrogen bombs floating aimlessly around in the void, fit them into our conceptual framework, test the degree to which they are manipulable, and then make our own plans, wherein an indefinitely sustained existence is seen only as a means and a good omen—but no guarantee—for the solution of problems that strike us as philosophically significant.

As I understand it, this is what took place on Zetetica. And—who knows?—did I or did I not originally fool myself into believing that I could persuade the Zeteticans to take my whole family to their planet? Then we could *really* have lived happily together ever after. I have never managed to admit to myself that I could be so naive. But who can say what wondrous, strange flowers of hope may grow in the darkest caverns of one's soul.

At any rate, whatever hope I unwittingly may have harbored for a Zetetican getaway has now most assuredly vanished, leaving not a trace behind. This is my end. *And my end is my end.*

But why end it here? And not in a bed in Hilary Hall, surrounded by my wife and children? They would all have been there: Taurus from Cambridge and Justus from Manchester—both of whom study astrophysics; Ina from the Academy of Art in Paris; and of course Teresa, who is still in high school. No, unfortunately, that was out of the question. At least, that is how I saw it. I shall explain why.

You may remember that I had been warned before my Tellurian return that the effect of the age immunization would slowly wear off when I resumed a human form of existence. I began to notice it four years ago: subtle signs of senility. I was sometimes unable to recall common words and names. Or I might utter a wrong word. It bewildered me. How often I wanted to enter into a conversation but then thought better of it for fear of not being able to finish my sentences. Sometimes I prepared sentences well in advance. And if there was a name in the sentence that I had a tendency to forget, I would lie in my bed for half an hour, repeating it over and over to myself.

I took every precaution to combat mental and physical enfeeblement. I jogged ten to twelve miles every morning and played tennis in the summer and handball during the winter. I took up cross-country skiing and ate health foods. I adopted a megavitamin diet. I even got hold of a special antisenility drug that had been invented for Generalissimo Franco of Spain.

I think it might have helped, at least for a while. My physician found me in excellent condition. But I did not.

I knew I was fighting a losing battle, and I felt humiliated when I faced myself: Had I not learned anything on Zetetica? Should I endure all this rigamarole merely to prolong my existence and add another brief allowance of breaths? Socrates found it unworthy to wait with the poison for another hour or so, although his friends assured him that "the sun was still upon the hills." Was not all this panicky activity and desperate pill popping also unworthy of me?

It was indeed a losing battle. Every day it got a bit worse. Few noticed it, I believe. Alicia, maybe, and possibly the boys. I could no longer take the same interest as before in quasars, pulsars, black holes, newly discovered gigantic galaxies, or even in the prospect of rewriting the whole scenario for the universe.

Reluctantly, I made up my mind: I would leave my family and let them remember me as I was.

The final deciding factor was furnished by a rather random and fortuitous affair. One of those social activities to which Alicia contributed so generously had to do with an "eventide home," as it was called. A sort of festivity was arranged to cheer up the old people, some of whom had children, grandchildren, and great-grandchildren visiting with them for the occasion. They also had hired a piano player, who had thick glasses and a complexion like a neglected colatorium. He was to accompany a huge blond entertainer, whom my father would have referred to as *"eine Gelee-Dame,"* * a plebeian howler in her fifties, callously looking over her audience of drivelling, babbling dotards. Some possessed ruddy, puffed-up,

---

*A "jelly-lady," directly translated; a woman, I suppose, with a rather high and gelatinous butterfat content.

paperish, dried-apple-peel faces and wet, pinkish eyes. Others were white, macerated, trembling, and transparent. One, completely paralyzed, was carried downstairs by some of his younger relatives. They stumbled, almost dropped him—and laughed secretly to each other. In the background lay another caducous case, supine, blowing into the air small cake crumbs that would land and stay stuck in his scrimpy grey beard, drenched by the colorless secretion from his nose. I saw flies dragging their feet in the slime. No one paid any attention. All eyes were on the *Gelee-Dame*. She addressed herself, with all the cheerfulness that a forty-pound honorarium could buy, to what she called "senior citizens" in their "golden years."

By contrast, a couple of middle-aged men at my left reviewed the concentration of anility in front of them with unconcealed disgust: "My God," one of them groaned, "are we going to end up like that? That boring? That repulsive?" "Sure as hell," replied the other mercilessly, "just you wait and see." "Not on your life," retorted the first. "I'll die first."

Suddenly the air was splintered by a stridor from the *Gelee-Dame*: "And now everybody chime in with me on the refrain: *'Happy days are here again!'*"

It made me physically sick to my stomach. I excused myself and ran back to Hilary Hall, leaving the car for Alicia. She picked me up after a few minutes. She was deeply concerned, but I brushed it off: "Probably that sweet punch."

But the next day—June 24—I did not put on my jogging clothes but dressed for hiking. I took a last look at Alicia. I could see from the movement of her eyes under her eyelids that she was dreaming. Sweet dreams, Alicia. Thank you for twenty-eight years of love and warmth and wise understanding. You know I wish it never had to end. I went from room to room without wakening any of the children, bidding them all silent farewells before I departed. Yes, I love them, perfectly knowing they are nothing but words out of the mouth of that same planet which I loathe and long to leave.

It was a lovely morning. I walked and walked, all the way to Amlwch, where I stayed overnight and caught a train the next morning to Newcastle, and from there a ship to Norway.

We had often hiked in the Norwegian mountains, Alicia and I, shortly after we married, and later, more infrequently, with the children. But never had we even come close to the little lake where the Zeteticans had picked me up and to which they returned me. And I had never mentioned a word about my temporary residence on Zetetica.

For nearly a month I wandered about on the mountain plateaus. Then, finally, I approached my lake again, this time from the mountains to the northeast. I could see the mountain dairy farm straight below me. It was deserted, not a living soul to see. Disappointed, I slowly descended. Strangely enough, there had obviously been people there, probably that very day. The grass had been cut recently and meticulously placed in the haydrying rack. It breathed a warm fragrance into my face. A rake, shining white, was resting against the rack. Spoken words still swam in the air. . . .

I slept that night in the same little hut where the three strangers had interrupted my attempt to translate Darwin's *Origin of Species* one hundred and fourteen summers ago. But the next day I moved on to my cottage. I preferred to remain undisturbed, even if there was hardly any food left. In fact, I had mostly vitamins and hormones and such useless things. But then, there were berries and trout. But I have long been too weak to go fishing. And the berries are rotting, yeasty, moldy.

So it goes.

October 3
TO MY READER*

Having eaten nothing at all for a week or so I feel lightheaded. No oxygen to speak of, I suppose. Or glucose. And no panic. Not even a mild concern: the blessed euthanasia of a bloodless, starving brain.

My dear reader, I have a favor to ask of you. I do not know, of course, who you are or whether you can read German in *Fraktur*. So I shall endeavor to make my request in English and with Latin letters.

I shall be dead when you find me. Please do not notify the authorities. There cannot be much left of me. It should be an easy task to leave my remains just anywhere out there where no one can find them. And, please, keep it your secret.

As for my diary, I wish I could have written it all in English. However, my vocabulary in that language is not what it was. Words refuse to come. But I should appreciate it if it could be translated some time. Tell the translator that he does not need to pay heed to my style or syntax. The only exceptions would be whenever I try to explain something I find particular-

*This last entry was written in English. I have made no changes.

ly difficult. And perhaps one more thing: I have taken pains to find the "right" words and tried to avoid the commonplace and platitudinous. I wish the translator would do the same.

I am vain enough to want my diary published. But whether that is possible or not, please remember to send the original to Mrs. Alicia Salcia von Habermann, Hilary Hall, Anglesey County, North Wales.

The sun is sitting low, shining straight into my eyes. It is warm and pleasant in here. *"Du grosses Gestirn, was ware dein Gluck wenn du nicht die hattest welchen du leuchtest!"** I have turned off the heater.

And just in case I should fail to awaken again, permit me only a last, little theatricality, a Faustian finale: *"Und die Erde hat mich wieder"*— "And the earth has received me back."

---

*A quotation from the beginning of Nietzsche's *Also Sprach Zarathustra:* "You great orb! What would your happiness be if you did not have those for whom you shine!"

Appendix:
For Further Reflection

Philosophers have always been known for two things: for knowing what they don't know, and for asking questions about it. Since Socrates demurred that he was the wisest man in all Greece because he was apparently the only man who knew what he didn't know, philosophers have asked more questions than they have answered, raised more problems than they have resolved.

It is in this spirit that this anthology is presented. In dealing with things biocosmic, there is no place for doctrinaire assumptions or simplistic conclusions. The ideas pondered by the authors of this volume are submitted for dialogue and further development by all who may find them of interest.

When this book was first projected, a series of philosophic questions, which we dubbed "Openers/Starters," was developed to give a modicum of direction to our thinking in this relatively new field. Those original questions are appended here in the hope that they might suggest other lines of speculation. They might initiate further dialogue or be used for course syllabi. They might stimulate analytical or empirical critique and papers. Or perhaps they just need to be pondered in a hundred directions until such time as the First Encounter occurs and we can rest content (or shudder forever) with the acquisition of definitive answers.

These questions are arranged in nine categories that are traditionally considered to be target areas for philosophical analysis:

1. Philosophy of religion
2. Philosophy of morality and ethics
3. Philosophy of communication and language
4. Philosophy of mind and consciousness (psychology plus)
5. Philosophy of esthetics
6. Philosophy of law systems and government
7. Philosophy of knowledge (epistemology plus)
8. Philosophy of death
9. The meaning of human existence in the context of a biocosmic ontology

1. THE PHILOSOPHY OF RELIGION

Planet earth is already inhabited by countless life-forms whose experiential worlds are undoubtedly alien to ours. We look at koalas, dung beetles, and dolphins with honest curiosity; but their impact upon our philosophic awareness is minimal because they are perceived as inferior or primitive by definition (man being the definer, of course). So we classify them and dismiss them—and read about them in *National Geographic*.

However, when we begin to speculate about the cosmos as we know it today, we face a new set of possibilities: that ETIs will inevitably be superior to man in significant and specific ways.

All present evidence indicates that ETIs exist, and if communication or encounter should occur, they will necessarily become "significant others" and will play crucial roles in the redefinition of man and the human condition. They will assume a paradigmatic status against which we will be forced to reassess the nature and significance of man.

It will be on the basis of the kind of philosophic ideas developed here that meaningful relationships with ETIs could proceed. Communication or contact will create a crisis for mankind—a crisis that, in its ultimate implications, is profoundly religious in nature.

Q Definition: What might the term *religion* possibly mean in a biocosmic perspective? Belief in deities? In spirits? In a supernatural order? A *concern* for ultimates? An involvement in ultimate *meaning?*

Q By any definition, what makes us think "religion" might be a universal phenomenon? Is it anything more than a human category of thought and feeling, or a term referring to a disparate cluster of human experiences?

Q Isn't religion merely part therapy, part myth, part security blanket, part fear of death, and part compensation for all-the-things-we-long-to-be-but-aren't?

One scholar writes, "Without alienation there is no religion because there is no *need* for religion." And an American Medical Association television commercial implores: "Attend the church of your choice, for faith is a physician." To be sure, alienation is correlated to many orthodox Western doctrines (atonement, redemption, salvation); and religion meets diverse human needs that are essentially therapeutic. Faith is found wherever system malfunctions occur.

Q Is alienation in fact the root of religion?

Q Is there good reason to believe that alienation (or something comparable) is universal?

Q What if we should have to relate to ETIs who have no such therapeutic needs?

Q How long would it take us to accept them (if ever)?

Q How long (if ever) would it take them to accept us?

Q Quite apart from all therapeutic needs, would ETIs still require "religion" to meet *other* needs?

Q Would the fact of death alone necessitate "religion"? (But what if they live for a thousand years or don't die at all? See Section 8, "The Philosophy of Death.")

Religion is closely linked to the way we value ourselves. Religious myths support and enhance our (lagging) feelings of self-esteem: man is created by God, in His image, a little lower than the angels. Human persons are sacred in the eyes of God. We have immortal souls (like the gods); we are, by divine intent, superior to animals; we are the objects of God's loving concern and the raison d'être of his plan of salvation.

Q How might we respond to ETIs who, perhaps inherently and naturally, possess strong feelings of self-worth?

Q Could we accept the implied mirror-image of ourselves?

Q What if the ETIs can't even understand man's problems to which these myths speak?

We have long used encompassing phrases such as "Ruler of the Universe" and "God of the Starry Sky"; and we have not infrequently conceived of a deity who occupied the totality of our world view.

Q On a cosmic scale, what might *god* or *deity* or *spirit* (and such terms with specific tribal and linguistic origins) mean?

Q Is such a notion as "cosmic deity" rationally conceivable?

Q What might happen to our traditional concepts and doctrines of God if—because man's collective self-image undergoes change—we can no longer "think small"? (Remember the theme of J. B. Phillips' book *Your God Is Too Small*.)

Q Can provincialism—even a planetary provincialism—survive in a biocosmos?

Q What will be the philosophic impact on man's feelings, beliefs, and ideas when he can no longer think anthropomorphically? (But philosophers don't think this way now, do they?)

> The Ethiopians make their gods black-skinned and snub-nosed; the Thracians say theirs have blue eyes and red hair. If oxen and horses had hands and could draw with their hands and make works of art just as men do, then horses would draw their gods to look like horses, and oxen like oxen—each would make their bodies in the image of their own.
>
> —Xenophanes

It has been a universal *human* trait to anthropomorphize—in good Disney-like fashion. If ETIs create cartoons, would they project their personal qualities onto animals, plants, deities, rocks, or whatever?

Q Can any living creature, in fact, escape the form of the "anthropomorphic fallacy"?

Q Aren't physical organisms everywhere subject to the limitations of the egocentric predicament, just as we are? If so, *how* might it be transcended?

Q Might some ETIs have progressed beyond "anthropomorphism" while allowing the masses (that is, other galactic beings, like us) to indulge their anthropomorphic needs?

Q Might they really tolerate our doing so?

Q Could *we* tolerate *their* doing so, especially if they turn out to resemble dung beetles revering that Great-Scarab-in-the-Sky—or something like that?

Q Would advanced ETIs have "need" of superfigures—omniscient authorities, all-forgiving mother/father deities, etc.—as we do?

Q Do these myths in fact represent transitional needs in the development of the genus Homo?

Q Is it conceivable that there is a universal stage in the development of higher life-forms comparable to our anthropomorphic phase?

For Aristotle, Aquinas, and several other Western philosophers, a supernatural figure was a metaphysical necessity: an Uncaused Cause, a Prime Mover, or the like.

Q In a biocosmic perspective, is the need for a Prime Mover resolved?

Q God is sometimes conceived of as the creator of matter, of the forces involving matter, and of cosmic order. In a biocosmic perspective, can such a concept be given meaning?

Q God is sometimes conceived of as the creator and sustainer of life. In a biocosmic perspective, can this concept be given meaning?

Q Are there possibly other forces—in addition to electromagnetic, gravitational, and nuclear forces—that might be cosmic in scope? Might they have "religious" significance?

Q Which of man's great religions would most readily accept the existence of ETIs?

Q Are some of man's religions already (perhaps implicitly) open to a biocosmos?

Science-fiction stories often tell of men who journey through the Galaxy as religious missionaries, usually with dire consequences to the "natives."

Q Might there be viable theological motives for human beings engaging in such cosmic missionary enterprises?

Q What conceivable rationales would justify ETIs evangelizing humankind? Might they muster religious zeal for belief systems, ways of life, survival principles, moral principles, political goals, military conquest, or mathematical truths?

2. THE PHILOSOPHY OF MORALITY AND ETHICS

With ETIs, we may for the first time have to work out ethical relationships with beings equal to or superior to us. It might be a whole new ethical ball game.

Q On what criteria might they be defined as worthy of our moral consideration?

Q On what criteria might they decide that *we* are worthy (or unworthy) of *their* moral consideration?

Q Would we necessarily assume that ETIs would have a serious "right to life"? Would ETIs necessarily be "persons" or "selves"?

Q Would the same evolutionary principles, from a moral perspective, hold true for ETIs as well as for us?

Q Might we discover that "survival of the fittest" in its bloodiest sense is indeed a universal morality?

Q Or might we discover that some form of "caring" is an ethical universal?

Q Might "fittest" (if applicable) mean "higher qualities": awareness, sensitivity, heightened consciousness, breadth of sensory range, rapid learning capacities? Might "fittest" really mean very advanced, complex qualities still wholly beyond us? Can we humans even imagine them?

When I was seven I enjoyed spending hours after school pounding big, red fire ants with a hammer. They scurried so fast and made elusive targets. I pounded thousands into the ground. I wasn't killing anything. They were wiggly objects that my parents regarded as a nuisance. They were not alive—I had never been told they were; they possessed no impulse to life. I had not been taught that they had sentient qualities; nor had I developed the capacity to fantasize and impute such qualities to them. Therefore, empathy was impossible for me at that stage.

Q What is the role of empathy in moral considerations? Would empathy (or its functional counterpart) be a universal quality?

Q Would we be "persons" to ETIs? Could they identify with us? Would our will to live be taken seriously? Should it be? What other needs or concerns might override their taking it seriously?

Q What if—because of their appearance—we couldn't empathize with them?

Q What if—because of our appearance—they couldn't empathize with us?

Q Is empathy learned or inherent? Apparently some human beings are entirely lacking in empathy for other human beings.

Mr. Spock of "Star Trek" is profoundly ethical. But wherein derives his ethics—from his logic or from his human side, which could *care* for others?

Q Can ethics derive from reason alone?

Q Can there exist a wholly rational morality?

Q If an organism's "caring capacities" are undeveloped, could there exist for it a genuine morality? (Or do such questions reduce to one more series of anthropomorphic fallacies?)

Q Can there be ethical relations without feelings?

Q Might feelings be universal qualities?

Sir Fred Hoyle is pessimistic about the survival of the human race for much longer. Hoyle is assuming (a) that all life-forms evolve through competition and the struggle to survive; (b) that evolutionary competition is necessarily cruel and violent; and (c) that there is, therefore, a stage of "brilliance plus destructiveness" that developing higher life-forms don't generally survive.

Q Are these assumptions sound? *Is* competition necessarily violent and destructive?

Q In fact, how sound are our reasons for believing that the "struggle for survival" is the *basic* universal evolutionary mechanism?

Q Does this struggle *necessarily* imply competition?

Q What about something else—say, "self-actualization"—as an evolutionary mechanism? That is, might "competition" be inner-directed rather than other-directed?

Q Might there be other ("unearthly") evolutionary mechanisms that would not only work for survival but would serve to produce ever greater biological/psychic complexity?

Imagine ETIs engaging in behavior that, to our way of thinking, is clearly immoral: ritual flagellation of themselves to develop "moral" qualities; indiscriminate sex with countless variations on a theme by de Sade; the practice of "emotional" violence to maintain "survival readiness"; and so on. The more humanoid they appeared to us, the more easily we would become threatened by behavior that, if we indulged in it, would be defined as "immoral."

Q What are our capacities for rapid redefinition?

Q What might be *their* capacities for rapid redefinition?

Q Could we develop a calculus of "minimal reaction" that would guide human behavior toward ETIs so that, as with other life-forms on earth, we would do the least possible harm to them?

Q Could we communicate such a calculus to them?

Q Would a biocosmic perspective serve to clarify the pragmatic character of moral codes?

A common sci-fi theme: Super-ETIs visit earth to thwart mankind's self-destruction.

Q Are there any conceivable reasons why ETIs might have any ethical concern for humans (or any other earth animals)?

Q Do *our* ethical concerns for other life-forms on earth give us any analogy for speculating on the preceding question?

Q What might be the effect on our collective self-image to take seriously the possibility that we are moral minims?

Q How might humans respond to a truly advanced or "higher" form of morality?

3. THE PHILOSOPHY OF COMMUNICATION/LANGUAGE

Simple abstractions—anything reducible to binary messages—will undoubtedly make up the content of the first communication with ETIs. And great amounts of information can be so communicated. But what we truly want to communicate and share are the deeper experiences that are the "stuff of experience."

Q Can we humans ever communicate *who* we are?

Q Do we in fact ever communicate *who* we are to one another?

Q If we do so at all, *how* do we do it? Might similar channels operate with ETIs?

Can *any* communication take place except on the basis of shared qualities? As a reigning principle, doesn't *absolutely all* communication of any kind build on the basis of characteristics that the communicators have in common? Therefore: Any communication with ETIs would derive from (a) universal qualities, plus (b) any qualities a specific ETI life-form happens to share with humans.

Communication between different life-forms would be facilitated if we had an accurate notion of universals. If a single ETI landed on earth tomorrow, what might we *assume* about him/it/her/whatever before we gathered a single item of information through empirical observation? That is, what possible universals might we imagine? Among others, could we *assume* the following?

He evolved in competition

He is in relationship

He has numerous survival qualities working for him

He can defend himself

He can perceive reality (or a coherent cluster of *some* realities)

He (or at least his species) can reproduce

He can conceptualize and think abstractly

He can communicate through one or more numerous possible media

He is endowed with sensory equipment that operates within the parameters of conditions that we can further speculate upon

He engages in valuing, selecting, and making choices

He engages in "goal-directed behavior"

He behaves in ways to escape pain (or an equivalent) and experience pleasure (or an equivalent)

He is a "person" and possesses a "self"

He has a "control module" (a CM—i.e., a brain equivalent) that transduces environmental energies into pragmatic images and concepts

He will have memory (some sort of information storage and retrieval system)

He will have a "need-hierarchy" or "need-priority" system

He will function in terms of some sorts of biological clocks

Or, does every one of these assumptions need critical analysis? For instance, what about the assumption of an evolutionary process and survival through competition? It is conceivable that ETIs might have been produced through genetic engineering, or that they may be the product of a controlled environment in which evolutionary competition as we know it plays no role whatever as a developmental mechanism. Also, how sound are our speculations that they would be capable of abstract thought and symbolic communication? What about emotional feelings? How many of these concepts are in fact just more anthropomorphic projections?

How could we ever communicate *experience* to ETIs if we humans can't communicate experience to one another? We are beginning to respect what we believe are different forms of intelligence in dolphins, octopuses, bees, and cats.

Q Would the problem with ETIs be basically the same—only more so?

Q What will be the impact on our self-image if we, not they, turn out to be the stupider of the communicators?

Q Might ETIs possess intelligence so different that the contents of their control modules ("brains") might come as a shock?

Q Might they possess such different valuing-systems that we can't comprehend them?

Q Might they behave in ways that offend us or merely escape all rationality as we know it?

Q Can we assume that all ETI forms of behavior will be intelligible once we understand the "reason" (i.e., motivation) behind it? Is it sound to assume that no conceivable ETI behavior is without meaning?

Pondering the problems of communicating with ETIs may throw into sharper relief the depth of the egocentric predicament not only of us humans but also of all organisms that occupy a physical body and are therefore limited to a self-point in space and time. This intolerable isolation could become too obvious; yet the painful loneliness of this ontological condition could hit us forcefully enough to make us do something about it. We might begin to explore the means of communicating experience now. It is even conceivable that biocosmic reflections on this problem could give us more immediate clarification and feedback.

Q Would transcendence of the egocentric predicament be a universal motivation behind attempts at communication? Would ETIs necessarily feel the frightening loneliness of physical isolation that characterizes the human condition?

Q Is there any conceivable hope of developing a means of communicating experience if symbols—all symbols—inherently and necessarily fail in this task?

4. THE PHILOSOPHY OF MIND/CONSCIOUSNESS (psychology plus)

Beyond and after electronic communication, direct knowledge of ETIs would have to begin through the observation of regularities in an organism's behavior.

Q Isn't this observation of matter-in-motion (the method of behaviorism) all that we are really doing in daily life in "reading" the behavior of others—humans as well as animals?

Q Or, do we in fact perceive "something more" than mere matter-in-motion when we are observing other *persons*?

Q Might the mere observation of behavior be a wholly inadequate method for understanding quite different forms of behavior?

Q Might we need to develop some type of "experiential psychology" to penetrate highly alien forms of experience?

In terms of "the philosophy of experience," if we are imaging all possible ETIs from every conceivable congenial environment, then some of them would undoubtedly embody forms of experience so strange and alien that it would be beyond our imagination.

Q Isn't this really what we already face in attempting (if we ever do) to understand the experience of nonhuman life-forms on earth?

Q As for ETI experience, would our problem merely be an extension of what we already know well? (Except that earthly parameters for possible experience are comparatively narrow; so in speculating on ETI experience we lose both the human and the larger earthly paradigms.)

ETI ecology: No ETIs could develop in isolation from specific environments or ecosystems. They will have evolved from primitive forms (probably) through struggle (probably) and in relationship to countless other competing life-forms (probably).

Q Wouldn't their behavior be intelligible to us only when we comprehend specifics about their complex system of relationships?

A sunflower "wants" (wills?) to keep its face turned toward the sun; but if it can't do so, then its "motivational goal" has been thwarted.

Q Can we be sure it doesn't experience pain, anger, emptiness, or sadness? Might it "feel" frustration?

Q Are there sound reasons for assuming the existence of emotions in all living creatures? Do plants have emotions? Do whales? robins? butterflies?

Q Might emotions (or their functional counterparts) be universals?

In advanced societies of ETIs, if basic survival needs have been met, what "self-actualizing" behavior might these fortunate organisms engage in?

Q What might their recreations be?

Q How might they utilize their leisure time?

Q Might they indulge in the production of art objects? If so, can we imagine what they might be like?

The "need hierarchies" for ETIs might be entirely different from man's. Fuel consumption and oxidation have top priority with us; and after that come shelter, sex, and other needs. But alien priorities could be quite different.

Q Wouldn't an accurate understanding of their need hierarchies be basic to our understanding of them and the establishment of relationships with them?

Q How might we go about understanding alien need hierarchies?

Q By what methods could we communicate to them *our* need hierarchies?

Are there any probable universals in the growth and development of organisms—e.g., infancy, adolescence, puberty, old age—in higher life-forms? (Again, earthly insect metamorphosis makes this question very difficult.)

Q Might we find forms of "neurosis" and "psychosis" or other control module (CM) malfunctions comparable to human mental and emotional aberrations? For instance, might there be a universal condition that (in man) we label "paranoia"? Wouldn't the same symptoms permit the same label?

Q Could we possibly develop universal classification systems for such disturbances? Don't we humans urgently need them now?

Q Wouldn't *all possible ETI forms* have the inherent potential for behavioral aberration?

Q Wouldn't CM malfunction ("brain damage") be a universal possibility?

Q What if CMs had the capacity for regeneration or repair (both physiologically and psychically)?

The mirror image: In a biocosmic perspective ETIs become a part of the mirror against which we define ourselves. (See Section 1.)

Q Would it be inevitable that we would come to see ourselves as ETIs see us, and define ourselves as they define us?

Q Is human personality *only* a mirror image? Are there other inherent or given elements in the development of human selves that are not other-dependent?

Q What if we are confronted by a very advanced ETI mirror?

Q In what ways might ETIs view us that could become significant in the reappraisal of our collective self-image?

Q What if the ETI mirror were, somehow, both inescapable *and* extreme to the point of being (for us) unacceptable or destructive?

Q Could we maintain our self-esteem in the face of strong ETI criticism and rejection?

Q If ETIs should proceed to "condition" us with reward and punishment, might we eventually become whatever they wish?

Q Might our encounter with advanced ETIs throw into relief *too much insight* into ourselves, for example, that genus Homo is presently in a painful, sadomasochistic, adolescent stage?

Q Would such insight be harmful or helpful to us?

5. THE PHILOSOPHY OF ESTHETICS

From *Planet of the Apes*

Taylor: [*To Zira*] Doctor, I'd like to kiss you goodbye.
Zira: [*Giggling*] All right. But you're so damned ugly!

De gustibus non disputandum est.
"Concerning taste there can be no argument."

If humankind ever needed an understanding of esthetic universals (if they exist) from which an objective axiology could be developed, it will surely be when we encounter ETIs; for at that moment all the puny differences between men will pale before esthetic shockwaves that could destroy all future relationships with ETIs.

In his *Philosophy of Art* Warren Steinkraus writes: "Westerners who make a fetish of cleanliness may not like to adorn their visages with the white ash of cow dung as some holy men of India do, but they cannot say that their preference is somehow objectively better than that of the sadhus." Yet "if it is possible to affirm anthropological laws transculturally, it might be possible to discuss, at least in a minimal way, aesthetic principles transculturally and objectively."

Q Isn't esthetic feeling largely, if not wholly, a matter of conditioned response? We find objects and events esthetically pleasing if they have been positively reinforced and displeasing if they are associated with negative reinforcement.

Q Won't our human esthetic responses show up as intolerably provincial when looked at carefully within a biocosmic world-view?

Q In terms of ETIs, doesn't the esthetic problem shape up largely as a problem of anthropomorphism? Desmond Morris cites a significant study in *The Naked Ape* that shows that our responses to other living species is positive or negative, strong or weak, in direct proportion to their humanlike appearance.

Q Or, do we respond primarily to what we perceive as compassionate behavior rather than to physical appearance? A 1969 study of Saturday-morning television cartoons showed that alienlike characters (e.g., Spider-man, Birdman, Vaporman) were considered friendly by children, and could be easily empathized with, since they "fought for Good and against Evil."

"To the lonely spaceman, an alien woman [with a mouth and nose vents like barn doors (designed for living in very thin atmospheres)] would hardly be attractive. Right here, the entire field of esthetics looms before us. Astronautical history may depend on those concepts of beauty and utility our men take along as unacknowledged cargo to the stars. Countless books will have to be written under the general title: *Esthetics and Etiquette for Other Worlds*. Otherwise, we are in danger of mistaking a rough skin for a rough mind, a third eye for an evil eye, a cold hand for a cold and hostile heart."   —Ray Bradbury

6. THE PHILOSOPHY OF LAW SYSTEMS AND GOVERNMENT

The basic purpose of law systems is to regulate behavior. Man's modus operandi has been to punish those who engage in any unacceptable behavior.

Q Doesn't the necessity of law systems for human beings mean that we possess freedom—i.e., freedom *from* internal behavioral determinants?

Q What if ETIs exist who—even though very advanced—possess a set of internal determinants (like ants or bees)?

Q Is this a strong possibility? Might they do what is considered right by "instinct" (where humans have apparently lost all instinct)?

Q Would laws for them therefore be irrelevant, superfluous, and meaningless?

Q Do ant colonies have "laws"? Do ants experience any "freedom"? Could intelligence develop *without* the loss of internal behavioral determinants?

Q Do freedom and law necessarily go together?

Q Can we assume that the following statements would apply universally? (a) The more competition and hostility, the more laws any society must have. (b) The more cooperation and caring, the fewer laws any society must have.

Laws are (or should be) entirely pragmatic; and they embody the values of the species and its social structure. Laws will be clues to what an ETI values and disvalues. Therefore, law systems among ETIs will vary enormously. But law systems for human beings reflect not merely freedom, but aberrations.

Q We assume ETIs will possess values, but what if ETIs experience no aberrations?

How might we respond if we discovered that ETIs were internally dominated by autonomous, self-actualizing drives—that is, that they were not "programmed" for conformist behavior but for loving, cooperative behavior? For them there would be no need for restraining laws but only for regulatory and predictive laws—guidelines, as in a chess game. Laws would apply to only a few aberrant individuals.

Q What does the fact that we have millions of laws say about us humans?

Laws and law systems (governments) occupy a central place in our existence when, ideally, they *should be* invisible. What kind of society will spend the major portion of its time and energy writing and rewriting the rules of the chess game but have no serious interests in playing the game or seeing it played? Hence, two apparent facts: (1) Crime ("aberrant behavior") has probably always been for human beings our greatest social problem. (2) Law-system development (politics)—*rather than productive actualization*—is a major preoccupation of our social lives.

Q What do these two facts (if indeed they are facts) say about the human condition (or the human "transition")?

Q Might some ETIs experience their collective existence in a wholly different way, i.e., might they be concerned with other activities than crime and politics and be free to devote almost full time to the pursuit of fulfilling goals?

Q Could we humans ever reach such a condition? What might preoccupy our time and energy then?

Q What might occupy such an extraterrestrial society now? What would be its goals, interests, causes, concerns?

A law *against* "murder" and "destructiveness": the very notion of such a law can sound weird—an anomaly, an absurdity—when pondered philosophically and biocosmically. Why? Why have a "law" against pounding your thumb with a hammer or against holding your hand over a fire? Are laws created for the sick? In fact, what is *government* but government of the sick, by the sick, and for the sick? (Apologies to Abraham Lincoln.)

Q How much of such a viewpoint is mere metaphor, how much is fallacy, and how much is fact or near fact? (We should come to some sort of answer before interpreting and evaluating the social systems of any other living creatures, shouldn't we?)

7. THE PHILOSOPHY OF KNOWLEDGE (epistemology plus)

Are the following statements basically sound? (A) Sensory range and dominance in all living organisms will develop in relationship to environmental factors. (B) CMs will organize perceptual data in pragmatic (survival) terms on a need-to-know principle.

Q Might these be universals? Or might there be other ways of sensing and conceptualizing—other "ways of thinking"? (Even within these two epistemological parameters, the mental construction of reality could take countless forms.)

We would normally expect the sensory range of ETIs to overlap that of humans to a limited extent and then to extend into realms beyond human perception (although not necessarily beyond the range of man's instruments). The chances of the sensory ranges of ETIs and humans coinciding exactly are very remote.

Human senses transduce energies emitted from a four-dimensional environment into neural signals that are utilized by the mind to construct dimensionless images and ideas with which we can make judgments about the real world. (Actually, they are one-dimensional constructs—that dimension being time—but I'll call them dimensionless for convenience.)

Q Can we assume that all intelligence will operate basically in this fashion, i.e., that the transduction of energies into pragmatic constructs is a universal phenomenon?

Q Or, might other CMs differ unimaginably in their mode of organization of perceptual raw materials?

Q Would "abstraction" be roughly the same process in all CMs? Incidentally, would all CMs necessarily be electrochemical systems?

Q Might there be energy forms (forces) that are unperceived or unorganized by us humans to which ETIs have adapted and developed sense receptors to respond to? What might such energy forms be? (Or is such a question so far beyond us that it must remain at present in the sci-fi domain?) Perhaps we have extended our sensory range with scientific instruments into normal ETI sensory range, and perhaps ETIs have extended their sensory range with instruments into what is our normal human sensory range.

Q In fact, isn't this inevitable? (See Section 3.)

Q Will ETI CMs organize information into patterns similar enough to ours to permit communication? How might their patterns differ?

Q Are there universals governing the organizational patterns of experience?

Q If evolutionary systems are teleological in any sense, might we find that there is some particular quality through which evolution is working toward "higher" life-forms? Might that quality be abstract reason, intuition, expanded consciousness, ESP or some mystical quality, greater "sanity" (reality perception), some emotional quality—or something entirely new and unknown to us?

8. THE PHILOSOPHY OF DEATH

There is reason to believe there will be striking differences in the ways ETIs perceive death, face death, experience death, interpret death, manipulate and control death; it is conceivable that some ETIs will not even experience death as we know it. In speculating biocosmically about death-events and death-practices generally, we are limited by the dullness of our imaginations and the earthly paradigms provided by living (and dying) organisms.

Suppose some ETIs have life-spans of three hundred years, or five hundred years, or a thousand.

Q Might we come to feel that man's life-span of three score years and ten is wholly intolerable?

Q How might we humans feel if ETIs are thereby enabled to fulfill a degree of their plans and dreams that has heretofore seemed unthinkable to us? Could we continue to accept docilely the early and frustrating curtailment of our dreams?

Q In terms of our self-esteem, could we continue to tolerate a seventy- to eighty-year life-span? Wouldn't we have to think of it as a sort of *premature death in infancy*?

George Bernard Shaw once said that if we humans lived for three hundred years we would be begging for death; for in three hundred years we would accumulate so much horror and misery that all of us would pray for the peace of nonexistence.

Similarly, Sir Fred Hoyle has written: "If I were given the choice of how long I should like to live with my present physical and mental equipment, I should decide on a good deal more than seventy years. But I doubt whether I should be wise to decide on more than three hundred years. Already I am very much aware of my own limitations, and I think three hundred years is as long as I should like to put up with them." But Hoyle goes on to imply that with an expansion of consciousness—one that would include, say, the consciousness of a Shakespeare, a Beethoven or Mozart, a Gauss, and more—perhaps a much longer lifetime would become desirable. The point is that one's life-span and the quality of that life are tied together.

Q Aren't these statements by Shaw and Hoyle essentially true in terms of man's present transitional stage?

Q But how long might we want to live *under present conditions* if we discover that some ETIs live for hundreds or thousands of our earth years?

Q If this might be the case, would we long even more for an immortal condition during which we could fulfill our dreams?

Q If an extraterrestrial species lives for a very long time, might we therefore assume that it has got things working better for it?

Q If we too could live for two thousand years, would our present profound need for immortality be affected? In what way?

In a biocosmic perspective, the fact of death as a cosmic event may be assessed and interpreted in new ways.

Q What grounds do we have for assuming that death-events as we know them are cosmic universals? Perhaps our notions are based solely on local and contingent examples.

Q What if some ETIs are merely casual about their deaths and accept termination with a quiet sense of completion?

Q "Rage, rage against the dying of the light." What if this attitude were assessed as intolerably neurotic by advanced ETIs? How might our human attitudes toward death be affected?

Q Is it remotely conceivable on our part that there may be forms of "process termination" other than death as we know it?

Q What is the probability that advanced ETIs will have developed methods of life-suspension as a normal part of their existence? (For purposes of space travel, some sort of life-suspension may be quite common.)

Q What might be some of the human reactions if it is indeed the case that various life-growth-death patterns exist?

Q Might some ETIs undergo a sort of metamorphosis?

Q Why has nature ("earth nature") chosen the four-stage metamorphic pattern (for example, as in butterflies) as a common developmental sequence? Does it have distinct survival characteristics? What are they? Might such characteristics serve higher life-forms also? (Or do we have butterflies only because they pollinate plants—that is, the form of the organism is produced to meet ecological requirements. Even so, of course, butterflies could be born directly from eggs.)

The problem of capriciousness in timing and the conditions in which human lives are terminated may turn out to be, in a biocosmic perspective, one of the most destructive and cruelest arrangements that living creatures could ever experience.

The pall and agony of the unknown—an all-pervasive cluster of anxieties that reduces man to a dysfunctional lump of matter that exits life stripped of the glorious dreams and accomplishments for which he worked a lifetime. At present, we can't even plan appropriate endings to richly fulfilled lifetimes!

Q Wouldn't higher life-forms at least attempt to establish control over both the time and the conditions of life-termination?

Q Wouldn't such control in fact be merely an extension of scientific and medical goals now on our human drawing boards?

Q In very advanced forms of ETIs wouldn't we normally expect to find almost complete control of death-events?

Q If such control could be accomplished by man, how might it affect the whole intent and meaning of human existence? (Would philosophers find themselves out of a job?) "When your number is up"—we have generally imputed to God or the Fates the choosing of the time and place of death. This capriciousness breeds endless varieties of rationales, fatalisms, and myths.

We can imagine two polar extremes regarding the value and control of death. On the one hand we can picture an authoritarian society in which the time and place of life-termination may be set by simple rules—rules that the individual ETI may know all his life and passively accept. On the other hand, in highly autonomous societies, individual ETIs may assume total and full control of the time and place for their own deaths. Having a death-event *happen to them* may be considered ignoble, tragic, or just plain messy. There may even be carefully designated times for life-termination.

Q Would either of these extremes be tolerable to us humans?

Q Would we find it possible to accept (without moralizing) ETIs who rigorously practice either extreme?

Q Would either of these extremes have any promise for human application? What might the benefits of either conceivably be?

Q What might be the human response to the ritual termination—perhaps even the ritual celebration of the termination—by an ETI of its existence?

Q Could we humans, at our present stage of development, accept such ritual termination for ourselves? Are there life-forms on earth that in fact practice such termination? Do we "accept" such a practice?

9. MEANING AND HUMAN EXISTENCE

Questions pondered here within the biocosmic perspective are not idle reflections. There is little doubt that if communication occurs with ETIs we humans will be forced into a position of having to relate to creatures beside whom we will not appear to be very bright or accomplished. And because they will be superior in numerous ways—many of which we probably won't understand—we will be in an insecure position and our existence will be fraught with anxiety until we can secure our relationships with them.

But the point is: *They exist now.*

The so-called real world is not the only thing that shapes what we are. More profoundly, we are molded by what we dream about, by our fantasies, by our visions and hopes for the future, by our pictures of better worlds and what might be. The very notion of a biocosmos opens up new worlds for us to live in—worlds that will indeed shape our daily existence profoundly.

We are molded and shaped by the worlds we live in. The man who lives in a small tribal world is a small tribal man. The man who lives in a black world or a white world is a black man or a white man. The man who lives in the world of the planet earth will be a planetary man. But the man who lives in the cosmos will be a cosmic man.

Men dream of other worlds because this world is not really habitable. There are many cosmic people who do not fit well into the world. But that doesn't require as much as it might seem, for this world is really very small.

Q Why would any living creature want to exist at all unless it experienced emotion? If existence doesn't feel good/bad/good—if there isn't pain (physical and emotional) to escape from and pleasure (physical and emotional) to seek out—would existence have any meaning at all?

Q Without feelings and senses is there anything else that would make an organism *want* to exist?

Q What would make a robot *want* to exist? "I want to exist. . . . I want to exist. . . . I am programmed to want to exist. . . . I want to exist. . . . " Does the robot really *want* to exist?

"If we can't think of something we want to do, then there's no reason to exist."

Bibliography:
For Further Reading

To date, the best philosophizing related to biocosmic questions has been carried on as a sort of side trip by the scientists who have worked deeply into biogenetic and extraterrestrial problems. One of Carl Sagan's virtues is that he humanizes his writing without compromising the hard science of his own discipline. His collaborative work with I. S. Shklovskii, *Intelligent Life in the Universe* (New York: Dell, 1966), remains a wellspring of philosophic ideas that pique our wonderment and become seed for thought.

A helpful, wide-ranging bibliography (with entries to 1974) is found in Cyril Ponnamperuma and A. G. W. Cameron, eds., *Interstellar Communication: Scientific Perspectives* (Boston: Houghton Mifflin, 1974), pages 187-226. The following publications should prove helpful in updating this basic bibliography and expanding it further into the realm of philosophy.

Ash, Brian. *Faces of the Future: The Lessons of Science Fiction.* New York: Taplinger, 1975.

Asimov, Isaac. *Is Anyone There?* New York: Doubleday, 1967.

Berendzen, Richard, ed. *Life Beyond Earth and the Mind of Man.* Washington, D.C.: NASA (Washington: U.S. Government Printing Office, SP-328), 1973.

Berry, Adrian. *The Next Ten Thousand Years: A Vision of Man's Future in the Universe.* New York: Saturday Review Press/E. P. Dutton, 1974.

Bracewell, Ronald N. *The Galactic Club: Intelligent Life in Outer Space.* San Francisco: W. H. Freeman, 1974.

Bradbury, Ray. "The Search for Extraterrestrial Life." *Life,* October 24, 1960.

Bundy, Robert, ed. *Images of the Future: The Twenty-First Century and Beyond.* Buffalo: Prometheus Books, 1976.

Dalzell, Bonnie. "Exotic Bestiary for Vicarious Space Voyagers." *Smithsonian* 5 (1974): 84-90.

Keosian, John. *The Origin of Life.* 2d ed. New York: Van Nostrand Reinhold, 1968.

Lovell, Bernard. *Man's Relation to the Universe.* San Francisco: W. H. Freeman, 1975.

Lunan, Duncan. *Interstellar Contact.* Chicago: Regnery, 1975.

Maruyama, Magoroh, and Arthur Harkins, eds. *Cultures Beyond the Earth: The Role of Anthropology in Outer Space.* New York: Vintage, 1975.

Miller, Stanley L., and Leslie E. Orgel. *The Origins of Life on the Earth.* Englewood Cliffs, N.J.: Prentice-Hall, 1974.

Moore, Patrick. *The Next Fifty Years in Space.* New York: Taplinger, 1976.

Ponnamperuma, Cyril. *The Origins of Life.* New York: E. P. Dutton, 1972.

Sagan, Carl, ed. *Communication with Extraterrestrial Intelligence.* Boston: MIT Press, 1973.

————. *The Cosmic Connection: An Extraterrestrial Perspective.* New York: Doubleday, 1973.

An abundance of avant-garde philosophic material related to biocosmic questions is to be found in science fiction. Many a sci-fi writer has pondered possible existential situations long and hard, or at least seen the problems clearly, and elaborated upon them in imagined real-life conditions. The following sci-fi selections are especially rich lodes of philosophic ideas relating to ETIs and man.

Asimov, Isaac. *I, Robot.* Gnome Press, 1950.

Bradbury, Ray. "Christus Apollo." *I Sing the Body Electric!* New York: Bantam, 1971.

Clarke, Arthur C. *Childhood's End.* New York: Ballantine, 1953.

————. *Rendezvous with Rama.* New York: Ballantine, 1974.

Heinlein, Robert A. *The Past Through Tomorrow.* New York: Berkley, 1975. (See "Methuselah's Children")

————. *Time Enough for Love.* New York: Berkley, 1974.

Herbert, Frank. *Dune.* New York: Ace Books, 1965.

Hollister, Bernard. *Another Tomorrow: A Science Fiction Anthology.* Dayton, Ohio: Pflaum, 1974.

Hollister, Bernard, and Deane C. Thompson. *Grokking the Future: Science Fiction in the Classroom.* Dayton, Ohio: Pflaum/Standard, 1973.

Keyes, Daniel. "Flowers for Algernon." In Isaac Asimov, ed., *The Hugo Winners,* vol. 1. New York: Doubleday, 1962. (This is the short story from which a full-length book and a motion picture, "Charly," were made. Though not dealing with ETIs, it focuses sharply on the problem of comparative intelligences.)

Lewis, C. S. *Out of the Silent Planet.* New York: Macmillan, 1965. (The most meaningful of the famous trilogy insofar as contact with ETIs is concerned.)

Stapledon, Olaf. *Last and First Men* (1931) and *Star Maker* (1937). Published in one volume. New York: Dover, 1968.

Zerwick, Chloe, and Harrison Brown. *The Cassiopeia Affair.* New York: Doubleday, 1968.